CURRENT ELECTRICITY

SK SAHDEV

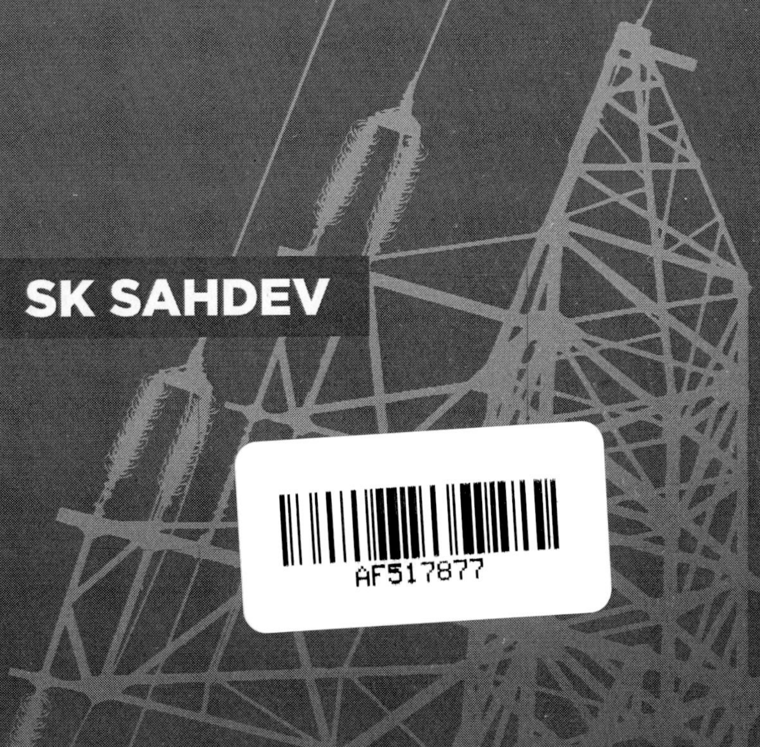

KHANNA BOOK PUBLISHING CO. (P) LTD.

Publisher of Engineering and Computer Books

4C/4344, Ansari Road, Darya Ganj, New Delhi-110002

Phone : 011-23244447-48 **Mobile:** +91-9910909320

E-mail : contact@khannabooks.com

Website : www.khannabooks.com

Dear Readers,

To prevent the piracy, this book is secured with HIGH SECURITY HOLOGRAM on the front title cover. In case you don't find the hologram on the front cover title, please write us to at contact@khannabooks.com or whatsapp us at +91-99109 09320 and avail special gift voucher for yourself.

Specimen of Hologram on front Cover title:

Moreover, there is a SPECIAL DISCOUNT COUPON for you with EVERY HOLOGRAM.

How to avail this SPECIAL DISCOUNT:

Step 1: Scratch the hologram
Step 2: Under the scratch area, your "coupon code" is available
Step 3: Logon to www.khannabooks.com
Step 4: Use your "coupon code" in the shopping cart and get your copy at a special discount
Step 5: Enjoy your reading!

Thanks,
Khanna Publishing House

Price: ₹299.00

ISBN: 978-93-90779-09-3

CURRENT ELECTRICITY
by **SK SAHDEV**

First Edition: 2025

To view complete list of books, please scan the QR Code:

Published by:
Khanna Book Publishing Co. (P) Ltd.
CIN: U22110DL1998PTC095547

Visit us at: www.khannabooks.com
Write us at: contact@khannabooks.com

Printed in Bharat.

Contents

CHAPTER

1 CURRENT ELECTRICITY

TOPICS TO BE DISCUSSED

Definition of Resistance, Voltage, Current, Power, Energy and their units, Relation between electrical, mechanical and thermal units, Temperature variation of resistance, Difference between AC and DC voltage and current

LEARNING OBJECTIVES

After the completion of this ***chapter,*** *the students/readers will be able to understand:*

1. Concept of electricity?
2. How Modern electron theory helps to understand electricity?
3. What is an electrically charged and neutral body?
4. Concept of electric potential and electric current?
5. What is resistance, resistivity, conductance and conductivity of a conductor and their units?
6. How emf is different to potential difference?
7. Ohm's law and its applications?
8. Effect of temperature on the resistance of metals, alloys, semiconductors and electrolytes.
9. How electric power is different to electrical energy?
10. What are the special units of energy when it is in the form of mechanical, electrical and heat?
11. Meaning of alternating voltage and current generally used commercially.
12. How to differentiate between ac and dc?'

INTRODUCTION

Electricity plays an important role in our day-to-day life. We are aware of the various functions that electricity can perform. It is used for lighting (lamps), heating (heaters), cooling (refrigerators), entertainment (radio and TV), transportation (electric traction), calculations (calculators) and many more. In fact, now – a – days, all our activities are totally dependent upon electricity. Because of this, it seems that most of the people are keen to know about electricity.

In this chapter, the readers will come across the basic concepts of currect electricity which enables them to understand the utility of electricity in the later stage.

1.1 ELECTRICITY

The invisible energy which constitutes the flow of electrons in a closed circuit to do work is called ***electricity.***

It is a form of energy which can be converted to any other form very easily. Electricity constitutes the flow of electrons in the circuit. It can be well explained by *Modern Electron Theory.*

1.2 NATURE OF ELECTRICITY

It has been seen that every matter is electrical in nature since it contains charged particles like protons and electrons. Therefore,

(*i*) ordinarily, a body is neutral as it contains same number of protons and electrons.

(*ii*) if some of the electrons are * removed from the body, there occurs a deficit of electrons and the body attains a *positive charge*.

(*iii*) if some of the electrons are supplied to the body, there occurs excess of electrons and the body attains a *negative charge*.

1.3 CHARGED BODY

Ordinarily, every substance or body is electricity neutral, as all the atoms of the body contain equal number of electrons and protons.

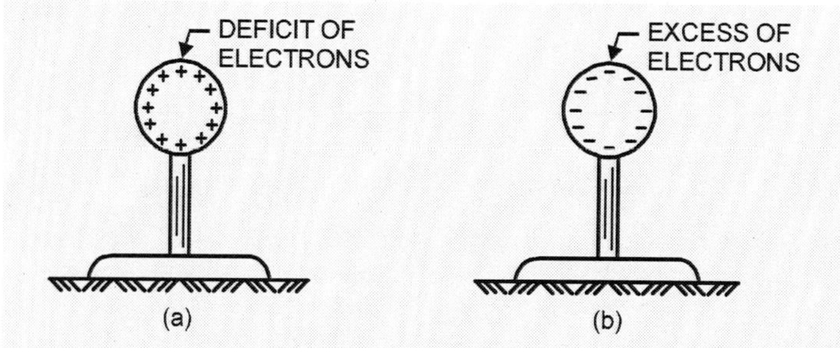

Fig. 1.1

However, when some electrons are detached from the body, there occurs a deficit of electrons form its normal due share. The result is that the body attains *positive charge* [see Fig. 1.1 (*a*)] where as, if electrons are supplied to a neutral body, there occurs an excess of electrons from its normal due share. The result is that the body attains *negative charge* [see Fig. 1.1 (*b*)].

Thus a body is said to be **charged** *positively or negatively if it has deficit or excess of electrons from its normal due share respectively.*

1.4 UNIT OF CHARGE

The *practical unit of charge is* **coulomb.**

1 coulomb = charge on 628×10^{16} electrons.

1.5 FREE ELECTRONS

The valance electrons which are very loosely attached to the nucleus of an atom and can be easily detached are called **free electrons.**

1.6 ELECTRIC POTENTIAL

The capacity of a charged body to do work is called **electric potential.**

The capacity of a charged body to do work determines the electric potential on it. Obviously, the measure of electric potential is the work done to charge a body to one coulomb, *i.e.*

$$\text{electric potential} = \frac{\text{Work done}}{\text{Charge}} \text{ or } V = \frac{W}{Q}$$

Unit: Since work done is measured in joule and charge in coulomb, the unit of electric potential is *joule/coulomb* or *volt.*

1.7 POTENTIAL DIFFERENCE

The difference in the electric potential of the two charged bodies is called **potential difference.**

Unit: The unit of potential difference is *volt.*

1.8 ELECTRIC CURRENT

When an electric potential difference is applied across the metallic wire, the loosely attached free electrons start drifting towards the positive terminal of the cell (see Fig. 1.3). This continuous drifting of electrons constitutes the electric current.

Thus, a continuous drifting of electrons in an electric circuit is called **electric current.**

The drifting of electrons in the wire is from *B* to *A i.e.* from negative terminal of the cell to the positive terminal through external circuit, whereas the conventional direction of flow of current is from *A* to *B i.e.* from +ve terminal of the cell to the –ve terminal through the external circuit.

The magnitude of flow of current at any section of the conductor is the rate of flow of electrons *i.e.* charge flowing per second.

Mathematically; $\text{Current, } I = \frac{Q}{t}$

Unit: Since charge is measured in coulomb and time in second, the unit of electric current is coulomb/sec (C/s) or ampere (A).

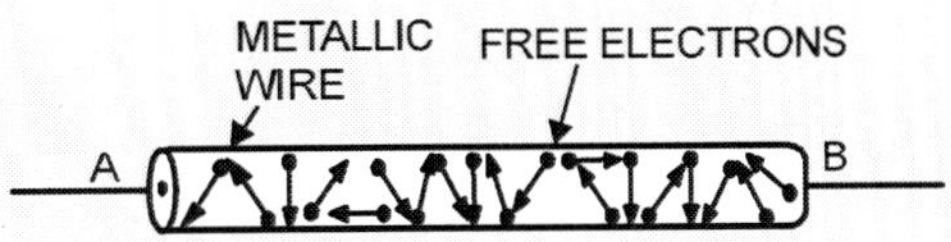

Fig. 1.2

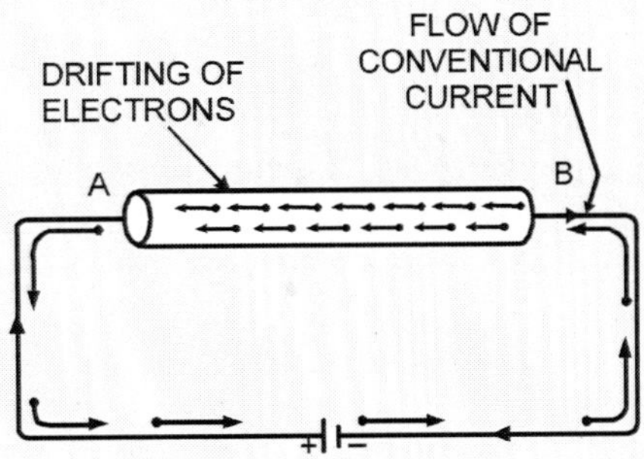

Fig. 1.3

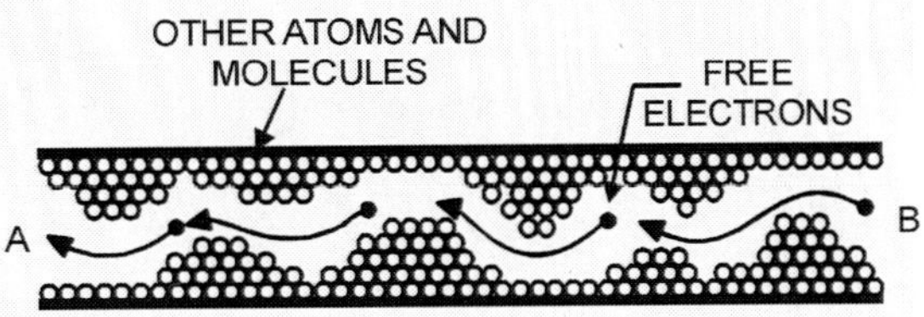

Fig. 1.4

1.9 RESISTANCE

The opposition offered to the flow of electric current (free electrons) is called **resistance.**

Unit: Resistance is measured in *ohm* (*or kilo-ohm*) and is denoted by symbol W (or kW).

1.10. LAWS OF RESISTANCE

The resistance (R) of a wire depends upon the various factors given below:

(*i*) It is directly proportional to its length, l *i.e.* R µ l

(*ii*) It is inversely proportional to its area of cross section, a *i.e.*

$$R \alpha \frac{l}{a}$$

(*iii*) It depends upon the nature (*i.e.* atomic structure) of the material of which the wire is made.

(*iv*) It also depends upon the temperature of the wire.

Neglecting the last factor for the time being; $R \alpha \frac{l}{a}$ or $R = \rho \frac{l}{a}$

Where r (**'Rho'** a greek letter) is a constant of proportionality called resistivity of the wire material.

1.11 RESISTIVITYM

The resistivity of a wire is given by the relation;

$$R = \rho \frac{l}{a}$$

If l = *1 m*; a = *1 m²* [*see Fig. 1.5 (a)*]; *then* $R = r$

Hence, the resistance offered by one metre length of wire of given material having an area of cross – section of one square metre is called the **resistivity** *of the wire material.*

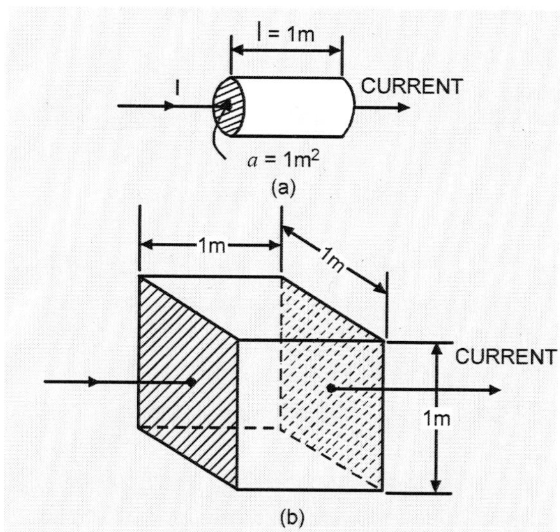

In place of wire, if a cube of one metre side of a given material is taken as shown in Fig. 1.5 (*b*). Considering opposite two faces of the cube.

$$l = 1 \text{ m}; a = 1 \times 1 = 1 \text{ m}^2 \text{ and } R = \rho$$

Hence, the resistance offered between the opposite two faces of one metre cube of the given material is called the **resistivity** *of that material.*

Unit: We know that $R = \rho \frac{l}{a}$ or $\rho = \frac{Ra}{l}$

Substituting the units of various quantities as per SI units, we get,

$$\rho = \frac{ohm \times m^2}{m} = \text{ohm} - \text{m}$$

Hence, the unit of resistivity is **ohm metre** *in SI units.*

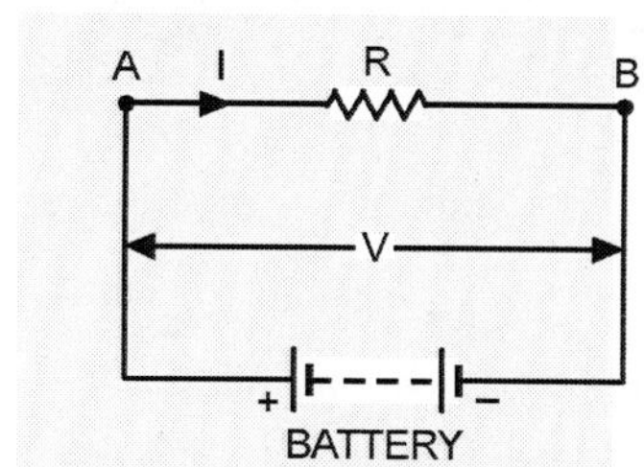

Fig. 1.6

1.12 SPECIFIC RESISTANCE

Specific resistance *of a material is defined as the resistance of the material having specific dimensions* *i.e.* one metre length and one square metre as area of cross section.

1.13 E.M.F. (ELECTROMOTIVE FORCE)

The **e.m.f.** *of a source, say a battery, is a measure of the energy that it gives to each coulomb of charge.*

At first sight, the name e.m.f. implies that it is a force that cause the electrons (the charged particles *i.e.* current) to flow through the circuit. In fact, it is not a force but it is energy supplied by some active source such as battery to one coulomb of charge.

1.14 E.M.F. AND POTENTIAL DIFFERENCE

The amount of energy supplied by the source to each coulomb of charge is known as **e.m.f.** *of the source. Whereas, the amount of energy used by one coulomb of charge in moving from one point to the other is known as* **potential difference** *between the two points.*

1.15 OHM'S LAW

Ohm's law *states that the current flowing between any two points of a conductor (or circuit) is directly proportional to the potential difference across them, provided physical conditions i.e. temperature etc. do not change.*

Mathematically;

$$I \propto V \text{ or } \frac{V}{I} = constant \text{ or } \frac{V_1}{I_1} = \frac{V_2}{I_2} = ... = \frac{V_n}{I_n} = \text{constant}$$

In other words, Ohm's law can also be stated as:

The ratio of potential difference across any two points of a conductor to the current flowing between them is always constant, provided the physical conditions *i.e.* temperature etc. do not change.

This constant is called the resistance (R) of the conductor (or circuit)

$$\therefore \qquad \frac{V}{I} = R$$

It can also be written as $V = IR$; $I = \frac{V}{R}$;

In a circuit, when current flows through a resistor, the p.d. across the resistor is known as *voltage drop* across it *i.e.* $V = IR$.

Limitations of Ohm's law: Ohm's law cannot be applied to the circuits:

(*i*) Consisting of electronic tubes or transistors because these elements are not bilateral.

(*ii*) Consisting of non-linear elements such as electric are etc.

Example 1.1. *The specific resistance of platinum at 0°C is 10.5 micro - ohm - cm. What should be the length of platinum wire of No. 32 S.W.G. (diameter = 0.0274 cm) to have a resistance of 4 ohms at 0°C ?*

Solution: Resistance of a wire at 0°C;

$$R_0 = \rho_0 \frac{l}{a} \quad \text{or} \quad l = \frac{R_0 a}{\rho_0}$$

where, $R_0 = 4$ W; $r_0 = 10.5 \times 10^{-6}$ W cm $= 10.5 \times 10^{-8}$ W m

$$a = \frac{\pi}{4}d^2 = \frac{\pi}{4}(0.0274)^2\,\text{cm}^2 = \frac{\pi}{4}(0.0274)\times 10^{-4}\ \text{m}^4$$

$$\therefore\quad l = \frac{4\times\pi\times(0\cdot0274)^2\times10^{-4}}{4\times10\cdot5\times10^{-8}} = 2.249\ \text{m}\ (Ans.)$$

Example 1.2. *A current of 0.75 A is passed through a coil of nichrome wire which has a cross - sectional area of 0.01 cm². If the resistivity of the nichrome is 108 × 10⁻⁶ ohm – cm and the p.d. across the ends of the coil is 81 volt, what is the length of the wire?*

Solution: Resistance, $R = \rho\frac{l}{a}$

Where, $R = \frac{V}{I} = \frac{81}{0\cdot75} = 108\ \Omega\ ;\ a = 0\cdot01\ \text{cm}^2 = 0\cdot01\times10^{-4}\ \text{m}^2$

$$\rho = 108\times10^{-6}\ \Omega\ \text{cm} = 108\times10^{-8}\ \Omega\ \text{m}$$

$$\therefore\quad l = \frac{V}{I} = \frac{Ra}{\rho} = \frac{108\times0\cdot01\times10^{-4}}{108\times10^{-8}} = 100\ \text{m}\ (Ans.)$$

1.16 EFFECT OF TEMPEREATURE ON RESISTANCE

The electrical resistance, generally changes with the change of temperature. The resistance does not only increase with the rise in temperature but it also decreases in some cases. In fact, the increase or decrease in resistance with the rise in temperature depends on the nature of the resistance material as discussed below:

(*i*) **Pure metals:** When the resistance is made of some pure metal (like copper, aluminium, silver etc.), its resistance *increases* with the *increase in temperature*. The increase is *large* and fairly *uniform* for normal range of temperature, therefore, temperature – resistance graph is a straight line. Thus, pure metals have *positive* temperature co – efficient of resistance.

(*ii*) **Alloys:** When the resistance is made of some alloy (liek Eureka, Manganin, Constantan etc.), its resistance *increases* with the *increase in temperature*. But the increase is very *small* and *irregular*. In the case of above said alloys, the increase in resistance is almost negligible over a considerable range of temperature.

(*iii*) **Semiconductors, insulators and electrolytes:** The resistance of semiconductors, insulators and electrolytes (like silicon, glass, varnish etc.) *decreases* with the *increase in temperature*. The decrease is non uniform. Thus, these materials have *negative* temperature co-efficient of resistance.

1.17 TEMPERATURE CO-EFFICIENT OF RESISTANCE

Consider a metallic resistor having a resistance of R_0 and 0°C and R_t at t°C. The increase in resistance $(R_t - R_0)$;

(*i*) is directly proportional to its initial resistance, *i.e.*

$$(R_t - R_0) \propto R_0$$

(*ii*) is directly proportional to rise in temperature, *i.e.*

$$(R_t - R_0) \propto t$$

(*iii*) depends on the nature of its material

Thus, $(R_t - R_0) \propto R_0 t$

or $(R_t - R_0) = \alpha_0 R_0 t$ ...(*i*)

where α_0 is a constant called *temperature co-efficient of resistance at 0°C*. Its value depends upon the nature of resistor material.

By rearranging the equation (*i*), we get,

$$R_t = R_0 (1 + \alpha_0 t) \quad ...(ii)$$

and
$$\alpha_0 = \frac{(R_t - R_0)}{R_0 t} \quad ...(iii)$$

If $R_0 = 1$ ohm; $t = 1$°C; Then, $\alpha_0 = (R_t - R_0)$

Hence, **temperature co-efficient of resistance at 0°C** *may be defined as the change in resistance per ohm original resistance per °C change in temperature.*

Unit: We know that, $\alpha_0 = (R_t - R_0)/R_0 t$

Substituting the units of various quantities, we get,

$$\text{Unit of } \alpha_0 = \frac{ohm}{ohm \times °C} = /°C$$

Hence, the unit of temperature co-efficient is *per °C*.

Example 1.3. *An aluminium resistor has a resistance of 43.6 ohm at 20°C and 50.8 ohm at 60°C. Determine the temperature co-efficient of resistance of aluminium at 0°C, its resistance at 0°C and at 40°C.*

Solution: $R_1 = R_0 (1 + \alpha_0 t_1)$ and $R_2 = R_0 (1 + \alpha_0 t_2)$

$$\therefore \quad \frac{R_2}{R_1} = \frac{1 + \alpha_0 t_2}{1 + \alpha_0 t_1} \text{ where } R_1 = 43.6\ \Omega;\ R_2 = 50.8\ \Omega$$

$$\frac{50.8}{43.6} = \frac{1 + \alpha_0 \times 60}{1 + \alpha_0 \times 20}$$

or $1600\ \alpha_0 = 7.2$ or $\alpha_0 = 0.0045/°C$ (*Ans.*)

Now,
$$R_0 = \frac{R_1}{1 + \alpha_0 t_1} = \frac{43.6}{1 + 0.0045 \times 20} = 40\ \Omega \text{ (Ans.)}$$

Resistance at 40°C, $R_3 = R_0 (1 + \alpha_0 t_3) = 40\ (1 + 0.0045 \times 40) = 47.2\ \Omega$ **(*Ans.*)**

Example 1.4. *A conductor has a cross-section of 8 cm² and specific resistance of 7.5 m ohm-cm at 0°C. What will be its resistance in ohm per kilometre when the temperature is 40°C ? Take temperature co-efficient of the material 0.005 per °C.*

Solution: Resistance of conductor at 0°C; $\rho_0 = \rho_0 \dfrac{l}{a}$

$$\text{Here, } \rho_0 = 7.5\ \mu\Omega - \text{cm} = 7.5 \times 10^{-8}\ \Omega\text{m};$$

$$l = 1 \text{ km} = 1000 \text{ m};$$
$$a = 8 \text{ cm}^2 = 8 \times 10^{-4} \text{ m}^2$$

$$\therefore \quad R_0 = \frac{7.5 \times 10^{-8} \times 1000}{8 \times 10^{-4}} = 0.09375 \text{ ohm}$$

Resistance of conductor at 40°*C*;

$$R_t = R_0 (1 + \alpha_0 t) = 0.09375 (1 + 0.005 \times 40) = \mathbf{0.1125 \text{ ohm}} \ (Ans.)$$

Example 1.5. *A copper coil has a resistance of 12.8 ohm at 20°C and 14.3 ohm at 50°C. Find (i) Temperature co-efficient of resistance at 0°C. (ii) Resistance of coil at 0°C. (iii) Resistance of coil at 100°C.*

Solution: $R_1 = R_0 (1 + a_0 t_1)$ and $R_2 = R_0 (1 + a_0 t_2)$

or $$\frac{14.3}{12.8} = \frac{1 + \alpha_0 \times 50}{1 + \alpha_0 \times 20}$$

or $$354 \alpha_0 = 1.5 \text{ or } \alpha_0 = \frac{\mathbf{1}}{\mathbf{236}} /°\mathbf{C} \ (Ans.)$$

Now $$14.3 = R_0 \left[1 + \left(\frac{1}{236}\right) \times 50\right]$$

or $$R_0 = \frac{14 \cdot 3 \times 236}{286} = \mathbf{11.8 \text{ ohm}} \ (Ans.)$$

Resistance of coil at 100°C, $R_3 = R_0 (1 + \alpha_0 t_3) = 11.8 \left(1 + \frac{1}{236} \times 100\right) = \mathbf{16.8 \text{ ohm}}$ (*Ans.*)

Example 1.6. *A motor winding has a resistance of 75 ohm at the room temperature of 20°C before switching ON to a 230V. After a 4 hour run the winding resistance is 90 ohm. Find the temperature rise if the material resistance temperature co-efficient is 1/234.5/°C.*

Solution: $a_0 = 1/234.5/°C = 0.004264/°C$

Resistance of winding at 20°C, $R_1 = 75$ ohm

Resistance of winding at t_2°C, $R_2 = 90$ ohm

$$\text{Now,} \quad \frac{R_2}{R_1} = \frac{R_0 (1 + \alpha_0 t_2)}{R_0 (1 + a_0 t_1)} \quad \text{or} \quad t_2 = 70.9°C$$

Rise in temperature = 70.9 - 20 = **50.9°C** (*Ans*)

TEST YOUR LEARNING-1

Type-I: Short Answer Questions

1. How will you define an electric charged body?
2. What is the unit of charge? Define one unit of charge.

3. What do you mean by electric potential and potential difference?
4. Define electric current and electric resistance.
5. What are the factors on which the resistance of a wire depends?
6. Define and explain ohm's law? What are its limitations.
7. What is the effect of temperature on pure metals, alloys, dielectric materials and semiconductors ?
8. Define temperature co-efficient of resistance, what are its units.

Type-II: Problems for Practice

1. What should be the cross - sectional area of a conductor of 1 km length to transmit 200 A so that the total voltage drop in the conductor may not exceed 12 volt. The resistivity of conductor material is 3 $\mu\ \Omega/cm^3$. *(Ans.* 5 cm^2)
2. A piece of $10 cm^3$ of copper having specific resistance of 1.7×10^{-6} ohm cm is (i) drawn into a wire of 200 m length. (ii) rolled into a square plate of 10 cm side. Determine the resistance of the wire and resistance between the opposite faces of the plate. *(Ans.* 60 Ω, 0.0017 $\mu\ \Omega$)
3. Find the resistance of a coil of mean diameter 4 cm containing 150 turns of manganin wire 0.05 cm in diameter. The resistivity of manganin is 42 μ ohm cm. *(Ans.*53.76 Ω)
4. There are two wires *A* and *B* of the same material. *A* is 20 times longer than *B* and has one fifth of the cross-section as that of *B*. If the resistance of *A* is one ohm, what is the resistance of *B* ? *(Ans.* 10 milli - ohm)
5. Find the resistance of 1500 m of a copper wire 25 square mm in cross-section. The resistance of copper is 1/58 ohm per metre length and 1 sq. mm cross section. What will be the resistance of another wire of the same material, three times as long and one half the cross-sectional area. *(Ans.* 1.0345 Ω, 6.207 Ω)
6. Two wires having equal length and made of the same material have resistance of 25 Ω and 81 Ω respectively. Find the ratio of their diameter. If the length of first wire is 50 cm and resistivity is $1.72 \times 10^{-8}\ \Omega$ m. Determine the diameter of each wire. *(Ans.* 1.8, 0.02093 mm, 0.01163 mm)
7. An aluminium resistor has a resistance of 43.6 ohm at 20°C and 47.2 ohm at 40°C. Determine the temperature co-efficient of resistance of aluminium at 0°C. *(Ans.* 0.0045/°C)
8. A conductor has a cross-section of 10 cm^2 and specific resistance of 7.5 μ ohm-cm at 0°C. What will be its resistance in ohm per kilometre when the temperature is 40°C? Take temperature co-efficient of the material 0.005 per °C. *(Ans.* 0.09 ohm)
9. A copper coil has a resistance of 12.7 ohms at 18°C and 14.3 ohms at 50°C. Find (i) Temperature co-efficient of resistance at 0°C. (ii) Resistance of coil at 0°C. (iii) Temperature co-efficient of resistance at 80°C. (iv) Resistance of coil at 100°C.

   (Ans. $\frac{1}{236}$/°C, 11.8 ohm, $\frac{1}{316}$/°C , 16.8 ohm)
10. A semi-circular ring of copper has an inner radius of 8 cm, radial thickness 6 cm and axial thickness 4 cm. Calculate the resistance of the ring at 80°C between its two end-faces. Specific resistance of copper at 20°C = 1.724×10^{-6} ohm cm. Resistance temperature co-efficient of copper at 0°C is 0.0043/°C. *(Ans.* 3.0716×10^{-6} ohm)

11. A motor winding has a resistance of 80 ohm at the room temperature of 20°C before switching *ON* to a 230V. After a 4 hour run the winding resistance is 100 ohm. Find the temperature rise if the material resistance temperature co-efficient is 1/234.5/°C. (*Ans.* 63.63°C)

12. A wire 100 cm long and 0.05 cm diameter is kept at room temperature 20°C. The specific resistance of the wire material is 3 micro-ohm per cm cube, while its temperature co-efficient of resistance is 0.004 at 20°C. Calculate the resistance of the wire at 20°C. If the wire is heated to 80°C, determine its increase in resistance. (*Ans.* 0.1528 Ω, 0.1894 ohm)

Type-III: Test Questions

1. Define the following and mention their units:

(*i*) Electric charge, (*ii*) electric current, (*iii*) electric potential (*iv*) emf (*v*) electric resistance (*vi*) conductance (*vii*) resistivity (*viii*) conductivity

2. State and explain Ohm's law. What are its limitations?

3. How temperature affects the resistance of metals, alloys and semiconductors. What is temperature co-efficient of resistance, what are its units?

4. How temperature affect the value of temperature co-efficient of resistance. Derive a relative

$$a_2 = \frac{1}{\frac{1}{\alpha_1} + (t_2 - t_1)}$$

1.18 ELECTRICAL ENERGY

When a potential difference (*V* volt) is applied across a circuit, as shown in Fig. 1.7, a current (*I* ampere) flows through it for a particular period (*t* second). A work is said to be done by the moving stream of electrons (or charge) and is called *electrical energy*.

Thus, the total amount of work done in an electrical circuit is called **electrical energy.**

By definition, $V = \frac{work\ done}{Q}$

Therefore work done or electrical energy expanded

$$= VQ = VIt \text{ (since } I = Q/t)$$

$$= I^2Rt = \frac{V^2}{R}t$$

where V = potential difference in volt;

I = in ampere;

t = time in second and

R = resistance in ohm.

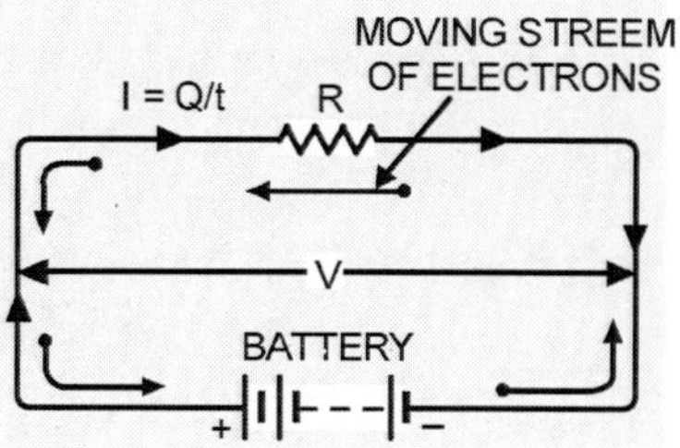

Fig. 1.7

Units: *The basic unit of electrical energy is* **joule (or Watt-second).**

If, V = 1 volt; I = 1 ampere and t = 1 second

Then, electrical energy = 1 joule

Hence, the energy expanded in an electrical circuit is said to be **one joule (or 1 watt-second)** *if one ampere current flows through the circuit for one second when a p.d. of 1 volt is applied across it.*

However, the practical or commercial unit of electrical energy is *kilowatt-hour* (*kWh*) which is also known as *B.O.T.* (*Board of Trade*) *unit.*

$$1\ kWh = 1000 \times 60 \times 60 \text{ watt-second} = 36 \times 10^5 \text{ Ws or joule}$$

Usually one kWh is called **one unit.**

1.19 ELECTRICAL POWER

The rate at which work is being done in an electrical circuit is called **electrical power.**

$$\text{Hence, electrical power} = \frac{\textit{Work done in an electric circuit}}{\textit{time}}$$

$$P = \frac{VIt}{t} = VI = I^2R = \frac{V^2}{R}$$

Unit: *The unit of electrical power is* **watt.**

If, $V = 1$ volt and $I = 1$ ampere. Then, $P = 1$ watt.

Thus, the power consumed in an electrical circuit is said to be **one watt** *if one ampere current flows through the circuit when a p.d. of 1 volt is applied across it.*

However, the bigger unit of electrical power is *kilowatt* (*kW*), it is usually used in the power system.

$$1 \text{ kW} = 1000 \text{ W}$$

1.20 MECHANICAL WORK

When a body, to which force is applied, moves in or opposite direction of the applied force, **work** *is said to be done by or against the body.*

Mathematically; Work = Force × distance or $W = F \times d$

Unit: *The unit of work is* **Newton – metre (Nm) or joule.**

If, $F = 1$ N and $d = 1$ m; Then, $W = 1$ Nm or joule.

Thus, when a force of *one newton* applied on the body moves it to a distance of *one metre*, the work done on the body is said to be *one newton - metre or joule.*

1.21 MECHANICAL POWER

The rate of doing work or the amount of work done per unit time is called **power,** *i.e.*

$$\text{Power} = \frac{\text{Work done}}{\text{time}}$$

Unit: *The unit of mechanical power is* **Newton metre per second** (*i.e.* Nm/s) *or* **joule/second** (*i.e.* **J/s**).

However, the practical unit of mechanical power is horse power.

In *fact, the rate of doing 75 kg m of work per second is known as* **one horse power.**

1.22 HEAT ENERGY

The form of energy which produces a sensation of warmth is called **heat.**

Mathematically;

Heat, $H = m\,S\,\theta$ Where, m = mass of the body; S = specific heat of the body; and θ = rise or fall in temperature.

Unit: *The unit of heat is* **kilocalorie (k cal)**

If, $m = 1$ kg; $\theta = 1°C$ and $S = 1$ *i.e.* specific heat of water. Then, H = 1 kilocalorie

Hence, *the amount of heat required to raise the temperature of 1 kg of water through 1°C is called* **one kilocalorie.**

However, the smaller unit of heat energy is *calorie.*

One calorie is defined as the amount of heat required to raise the temperature of 1 gram of water through 1°C.

1 kilocalorie = 1000 calorie

1.23 JOULES LAW OF ELECTRICAL HEATING

Joule (James Prescott Joule) established that there exists a definite relation between electrical energy expended and amount of heat produced. Thus, the relation is called Joule's law of electrical heating.

This law stated that *the amount of heat produced (H) is directly proportional to the electrical energy expended (W).*

i.e. $$H \propto W \quad \text{or} \quad \frac{W}{H} = J\,(\text{constant})$$

Where J is a constant called mechanical equivalent of heat and its value is determined as *4.18 joule per calorie* (*i.e.* 1 calorie = 4.18 joule). It means that to produce one calorie of heat, 4.18 joule of electrical energy is expended.

From equation (*i*), we get, $$H = \frac{W}{J} \text{ or } H = \frac{I^2 Rt}{4.18} \text{ calorie}$$

where $I^2 Rt$ is the electrical energy in joule.

1.24 RELATION BETWEEN VARIOUS QUANTITIES

Some of the important relations between various electrical, mechanical and thermal (heat) quantities are given below:

1. Relation between horse power and kW

By definition; 1 H.P. = 75 kg - wt m/s

= 75 × 9.81 Nm/s or joule/s or watt

= 735.5 W *i.e* **1 H.P. = 0.7355 kW**

2. Relation between horse power and torque

If a rotor of a radius r metre rotates at a speed of N r.p.m. The force acting on the rotor tangential to its radius is F newton, then

$$\text{Work done in one rotation} = \text{Force} \times \text{distance covered/rev}$$
$$= F \times 2\pi r = 2\pi T \text{ Nm or joule}$$

where T is the torque *i.e.* moment acting on the rotor.

$$\text{Work done / minute} = 2\pi NT \text{ (since } N \text{ revolutions are made in one minute)}$$

$$\text{Work done / sec or power} = \frac{2\pi NT}{60} \text{ joule/s or watt}$$

$$\therefore \quad \textbf{H.P.} = \frac{\mathbf{2\pi NT}}{\mathbf{60 \times 735 \cdot 5}} \text{ (because 1 H.P. = 735.5 W)}$$

3. Relation between kWh and k calorie.

Since
$$1 \text{ kWh} = 1000 \times 60 \times 60 \text{ Ws or joule} = \frac{36 \times 10^5}{4 \cdot 18} \text{ calorie}$$

$$= \frac{36 \times 10^5}{4 \cdot 18 \times 1000} \text{ k cal}$$

$$\therefore \quad \mathbf{1 \text{ kWh} = 860 \text{ k cal}}$$

Example 1.7: *A building has (i) 12 light points of 60 watt each burning 4 hour a day, (ii) a fan point of 75 watt each running 10 hour a day, (iii) a plug point for a 750 watt heater is used one hour a day. (iv) one radio 80 watt used 6 hour a day and (v) a 1/2 H.P. pump of 80% efficiency running 2 hour a day. Calculate the total connected load in kilowatt, maximum possible current, the daily consumption of energy and monthly bill. The supply is given at 230 volt and energy costs Rs. 5.15 per unit. The rent for a meter is Rs. 50 per month. Assume the month of 30 days.*

Solution:

Load Points	Connected Load	Energy Consumed/Day
(*i*) 12 lights of 60W each 4 hr/day	12 × 60 = 720 W	$\frac{720 \times 4}{1000} = 2.88$ kWh
(*ii*) 4 fan points of 75W each – 10 hr/day	4 × 75 = 300 W	$\frac{300 \times 10}{1000} = 3.00$ kWh
(*iii*) 1 plug point of 750W heater – 1 hr/day	1 × 750 = 750W	$\frac{750 \times 1}{100} = 0 \cdot 75$ kWh
(*iv*) 1 ratio of 80W – 6 hr/day	1 × 80 = 80 W	$\frac{80 \times 6}{1000} = 0 \cdot 48$ kWh
(*v*) 1/2 H.P. Pump 80% eff. – 2 hr/day	$\frac{1}{2} \times \frac{735 \cdot 5}{80} \times 100 = 460$ W	$\frac{460 \times 2}{1000} = 0 \cdot 92$ kWh
Total	**2310 W**	**8.03 kWh**

Therefore connected load, $P = 2310 \text{ W} = \mathbf{2.31 \text{ kW}}$ (*Ans.*)
Maximum possible current, $I = P/V = 2310/230 = \mathbf{10.043 \text{ A}}$ (*Ans.*)
Energy consumption/day $= \mathbf{8.03 \text{ kWh}}$ (*Ans.*)
Energy consumption/month $= 8.03 \times 30 = \mathbf{240.9 \text{ kWh}}$ (*Ans.*)
Rate of energy/month = Rs. 5.15
Energy cost/month $= 5.15 \times 240.9 =$ Rs. 1240.60
Meter rent/month = Rs. 50.00
Therefore monthly bill $= 50 + 1240.60 =$ **Rs. 1290.60** (*Ans.*)

Example 1.8. *An electric kettle was marked 500 watt, 230 volt and was found to take 13 minute to bring 1 kg of water at 20°C to boiling point. Determine the heat efficiency of the kettle.*

Solution: Heat absorbed by water *i.e.* output of kettle,

$$H = m\,S\,\theta$$

where $m = 1\text{kg} = 1000 \text{ g};\ S = 1;\ \theta = t_2 - t_1 = 100 - 20 = 80°\text{C}$

$$\therefore \quad H = 1000 \times 1 \times 80 = 80000 \text{ calorie}$$

Energy input to kettle = Power × time

$= 500 \times 13 \times 60 = 390000$ watt - sec or joule

$= 390000/4.18$ calorie $= 93301$ calorie

$$\text{Heat efficiency of kettle} = \frac{\text{Heat utilised by water}}{\text{Heat produced in kettle}} = \frac{80000}{93301} = 85.74\% \ (\textit{Ans.})$$

Example 1.9. *A geyser heater rated at 3 kW is used to heat its copper tank weighing 20 kg and holds 80 litter of water. How long will it take to raise the temperature of water from 10°C to 60°C, if 20% of energy supplied is wasted in heat losses.*

Assuming specific heat of copper to be 0.095 and 4.2 joule to be equivalent to one calorie.

Solution:

Mass of water, $m_1 = 80$ kg (since 1 litter of water weighs 1 kg)
Mass of tank, $m_2 = 20$ kg;
Specific heat of copper, $S_2 = 0.095$
Change in temperature, $\theta = (t_2 - t_1) = 60 - 10 = 50°\text{C}$

Heat utilised to raise the temperature of water and tank or output

$$= m_1 S_1 \theta + m_2 S_2 \theta = 80 \times 1 \times 50 + 20 \times 0.095 \times 50$$

$$= 4095 \text{ k cal} = 4095 \times 10^3 \times 4.2 \text{ joule}$$

Thermal efficiency, $\eta = 80\%$ (since loss = 20%)

$$\therefore \quad \text{Input energy} = \frac{\text{output}}{\eta} = \frac{4095 \times 10^3 \times 4.2}{0.8} \text{ joule}$$

$$= \frac{4095 \times 10^3 \times 4.2}{0 \times 8 \times 36 \times 10^5} \text{ kWh} = 5.973 \text{ kWh}$$

Time required to raise the temperature.

$$= \frac{\text{Input energy}}{\text{power}} = \frac{5.973}{3} = \textbf{1.991 hour } (Ans.)$$

Example 1.10. *A diesel electric generating set supplies an output of 100 kW. The calorific value of the fuel oil used is 12500 k cal/kg. If overall efficiency of the unit is 36%, (i) calculate the mass of oil required per hour and (ii) the electrical energy generated per tonne of the fuel.*

Solution: *Calorific value:* The heat produced by the complete combustion of one kg of fuel is called the calorific value of the fuel.

$$\text{Output} = 100 \text{ kW}$$

$$\text{Energy delivered/hour} = 100 \times 1 = 100 \text{ kWh}$$

$$\text{Energy input} = \frac{\text{output}}{\eta} = \frac{100 \times 100}{36} = 277.78 \text{ kWh}$$

$$\text{Heat energy required} = 277.78 \times 860 \text{ k cal} = 238889 \text{ k cal}$$

$$\text{Fuel required/hour} = \frac{\text{Heat produced}}{\text{calorific value of the fuel}} = \frac{238889}{12500} = \textbf{19.11 kg } (Ans.)$$

$$\text{Heat produced/tonne of fuel} = 12500 \times 1000 \text{ k cal}$$

$$\text{Electrical energy generated} = \frac{\textit{Heat produced}}{860} \times \eta = \frac{12500 \times 1000}{860} \times \frac{36}{100}$$

$$= \textbf{5232.56 kWh } (Ans.)$$

Example 1.11. *A hydro - electric power station operates at a mean heat of 25 m and is supplied from a reservoir having area of 6 sq. km. Calculate the energy produced by the water if the water level in the reservoir decreases by 1 m. The overall efficiency of the power station may be considered 80%.*

Solution:

$$\text{Area of reservoir} = 6 \text{ km}^2 = 6 \times 10^6 \text{ m}^2$$

$$\text{Decrease in water level} = 1 \text{ m}$$

$$\text{Volume of water used} = 6 \times 10^6 \times 1 = 6 \times 10^6 \text{ m}^3$$

$$\text{Mass of water, } m = 6 \times 10^6 \times 1000 \text{ kg (1 m}^3 \text{ of water weighs 1000 kg)}$$

$$\text{Height of water fall or head, } H = 25 \text{ m}$$

$$\text{Potential energy of water fall} = mgH$$

$$= 6 \times 10^9 \times 9.81 \times 25 \ (\_\ g = 9.81)$$

$$= 14\,715 \times 10^8 \text{ Nm}$$

Energy utilised to generate electrical energy, *i.e.*

$$\text{Output} = \text{Input} \times \eta = \frac{14715 \times 10^8 \times 80}{100} \text{Nm or joules}$$

$$= \frac{14715 \times 10^8 \times 80}{100 \times 1000 \times 60 \times 60} = \textbf{327000 kWh } (Ans.)$$

TEST YOUR LEARNING-2

Type-I: Short Answer Questions

1. Define electrical energy and electric power, what are their units?
2. What do you meen by HP of the motor?
3. Derive a relation between (*i*) electrical energy and heat energy, (*ii*) electric energy and mechanical energy.

Type-II: Problems for Practice

1. An electric boiler draws 12 ampere at 115 volt for a period of 6 hour; (i) If electrical energy costs Rs. 3.00 per kilowatt hour, determine the cost of boiler operation. (ii) Find the quantity of electricity in coulomb passing through the boiler. (***Ans.*** Rs. 24.84, 259200 C)
2. The demand for the lightning of a small village is 50 A at 200 V. This is supplied from a dynamo at a distant station. The generated voltage being 220 V. Find (i) The resistance of the leads from the dynamo to the village. (ii) The number of BOT units of energy (kWh) consumed in 10 hours in the village. (iii) number of BOT units of energy (kWh) wasted in the leads in the same time. (***Ans.*** 0.4 Ω, 100 BOT units, 10 BOT units)
3. The potential difference of 10 V is applied across a 2.5 ohm resistor. Calculate the current, the power dissipated and energy consumed in heat units in 5 minute. (***Ans.*** 4 A, 40 W, 2870 calorie)
4. An electric kettle was marked 500 watt, 230 volt and was found to take 15 minute to bring 1 kg of water at 15°C to boiling point. Determine the heat efficiency of the kettle. (***Ans.*** 78.95%)
5. An immersion heater rated at 3 kW is used to heat a copper tank weighing 20 kg and holds 120 litter of water. How long will it take to raise the temperature of water from 10°C to 60°C, if 20% of energy supplied is wasted in heat losses.

   Assuming specific heat of copper to be 0.095 and 4.2 joule to be equivalent to one calorie. (***Ans.*** 2.963 hour)
6. A diesel electrical generating set supplies an output of 50 kW. The calorific value of the fuel oil used is 12500 k cal/kg. If overall efficiency of the unit is 36%, (i) calculate the mass of oil required per hour and (ii) the electrical energy generated per tonne of the fuel. (***Ans.*** 9.55 kg, 5232.56 kWh)
7. A hydro-electric power station operates at a mean heat of 30 m and is supplied from a reservoir having area of 5 sq. km. Calculate the energy produced by the water if the water level in the reservoir decreases by 1 m. The overall efficiency of the power station may be considered 80%. (***Ans.*** 3 27 000 kWh)

Type-III: Test Questions

1. Derive a relation between (*i*) HP and watt (*ii*) kWh and kCal (*iii*) HP and torque.

1.25 ALTERNATING VOLTAGE AND CURRENT

A voltage that changes its polarity and magnitude at regular intervals of time is called an *alternating voltage*.

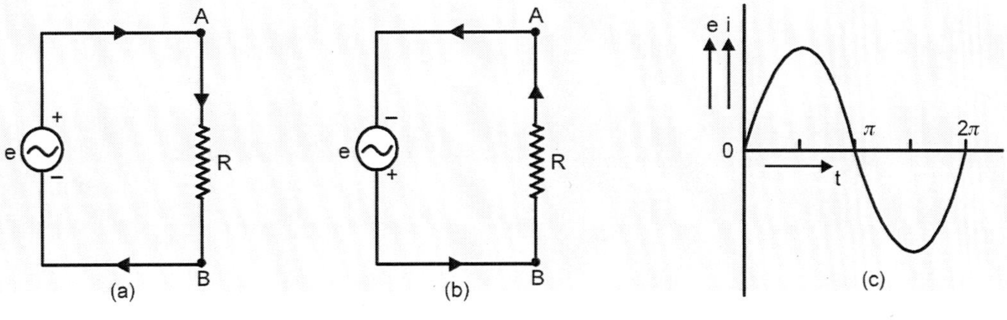

Fig. 1.8

When an alternating voltage source is connected across a load resistor *R*, as shown in Fig. 1.8, the current flows through it in one direction and then in opposite direction when the polarity is reversed.

Fig. 1.8 (*c*) shows the wave shape of the source voltage (representing the variation of voltage w.r.t. time) and current flowing through the circuit (*i.e.* load resistor *R*).

Wave form: The graph representing the manner in which an alternating voltage or current changes w.r.t. time is known as *wave-form* or *wave-shape*. While plotting a graph, usually the instantaneous values of the alternating quantities are taken along *y-axis* and time along *x-axis*. The alternating voltage or current may vary in different manner, as shown in Fig. 1.9, accordingly their wave shapes are named in different ways such as irregular wave; triangular wave; square wave; periodic wave; saw-tooth wave; sine wave etc.

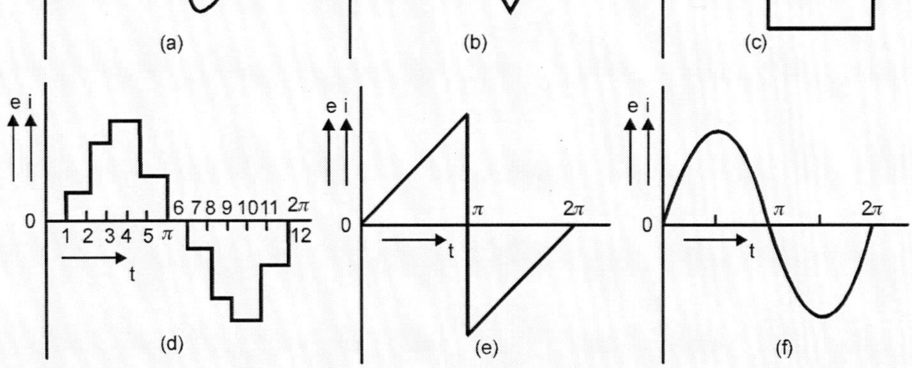

Fig. 1.9

1.26 DIFFERENCE BETWEEN AC AND DC

AC current	DC current
1. Alternating current reverses periodically and its magnitude changes.	1. Direct current flows only in one direction and remains unaltered.
2. Amplitude and polarities are varying continuously.	2. Amplitude and polarities are fixed.
3. It has a particular frequency.	3. It is independent of frequency.
4. *AC* can be generated at higher voltages.	4. *DC* cannot be generated at high voltages because of commutation difficulties.
5. In case of *ac*, the cost of generation is less.	5. In case of *dc*, the cost of generation is more.
6. Alternating voltage can be increased (stepped up) or decreased (stepped down) easily with the help of a transformer.	6. Direction voltage cannot be increased or decreased easily.
7. *AC* motors are of less cost, more robust and durable.	7. *DC* motors are costly and less durable.
8. The maintenance cost of ac equipment and appliances is less.	8. The maintenance cost of *dc* equipment and appliances is more.
9. *AC* cannot be used directly for electroplating.	9. Only *dc* can be used directly for electroplating.
10. The speed of *ac* motors cannot be controlled easily.	10. The speed control of *dc* motor is very easy and economical.

SUMMARY

1. *Current*: The magnitude of flow of current at any section of the conductor is defined as the rate of flow of charge (electrons) at that section, $I = \frac{Q}{t}$ C/s or ampere.
2. *Laws of resistance:* It is summed up as $R = \rho\frac{l}{a}$ ohm
3. *Ohm's law*: Ratio of V to I is constant for a circuit provided physical conditions remain the same $R = \frac{V}{I}$; $V = IR$ or $I = \frac{V}{R}$
4. *Temperature co-efficient of resistance at 0°C:* $\alpha_0 = \frac{(R_t - R_0)}{R_0 t}/°C$
5. *Effect of temperature on resistance (for conductors):* Resistance increases proportionately with the rise in temperature $R_t = R_0\,(1 + a_0 t)$; $R_{t_2} = R_{t_1}\,[1 + \alpha_{t_1}\,(t_2 - t_1)]$
6. *Electrical energy*: When current flows through a circuit, an electrical work is being done, called electrical energy.
   Electrical energy: VIt josule or watt-sec; Commercial unit or *BOT* unit is kWh; $1kWh = 36 \times 10^5$ joule.

7. *Electrical power:* Electrical energy consumed per unit time is called *electric power* ($VIt/t = VI$).
8. *Relation between various quantities:*

   1 Calorie = 4.18 joule; 1 k cal = 1000 calorie = 4186 J; 1 kWh = 860 k cal

   1 H.P. = 735.5 W = 0.7335 kW; $H.P. = \frac{2\pi NT}{60 \times 735 \cdot 5}$ (T is in Nm)
9. *Electrical power:* Electrical energy consumed per unit time is called *electric power* ($VIt/t = VI$).
10. *Alternating current:* The current that chnages in magnitude and direction periodically is called alternating current.

SELF ASSESMENT

TEST YOUR PREPARATION

OBJECTIVE TYPE QUESTIONS

Type-I: Fill in the blanks:

1. Resistance of a conductor is __________ proportional to its area of cross-section.
2. The resistance of an insulator __________ with the increase in its temperature.
3. A 3/20 S.W.G. wire can carry __________ current than 7/20 S.W.G. wire.
4. kWh is the unit of __________.
5. One unit of electrical energy means __________.
6. The resistance of 40 W, 200 V lamp will be __________ than 100 W, 200 V lamp.
7. An electric lamp of 40 W will consume one unit of energy in __________ hour.
8. The resistance of a conductor __________ with the decrease in temperature.
9. Ohm's law will hold good if the physical conditions particularly __________ remains constant.
10. The resistance of a conductor __________ with the decrease of its diameter.
11. The electric charge on an electron is __________ coulomb.
12. One coulomb of charge is equal to the charge on __________ electrons.
13. A 100 W, 220 V lamp will take __________ ampere current at rated voltage.
14. Watt hour is the unit of __________.
15. Energy consumed by a 750 W, 230 V electric iron in 8 hour will be __________ unit.

Type-II: Multi-choice questions:

1. *Electrons which are loosely attached to the nucleus of an atom and can be easily detached are called*

   (*a*) free electrons (*b*) valance electrons

   (*c*) bonded electrons (*d*) all *a*, *b* and *c*

2. *Ohm's law deals with the relation between*
(*a*) charge and resistance (*b*) charge and capacity
(*c*) current and p.d. (*d*) capacity and p.d.

3. *If the length of a conductor or wire is doubled and its area of cross section is also doubled, then the resistance will*
(*a*) increase four times (*b*) remains uncharged
(*c*) decrease to four times (*d*) change at random

4. *The unit of electrical energy is*
(*a*) watt -sec (*b*) joule (*c*) kWh (*d*) all *a*, *b* and *c*

5. *A 100 watt, 220 volt lamp takes a current of*
(*a*) 5/11 A (*b*) 2.2 A (*c*) 0.22 A (*d*) 5/22 A

6. *Watt hour is the unit of*
(*a*) electric power (*b*) electric capacity
(*c*) electric energy (*d*) electric charge

7. *One kWh is equal to*
(*a*) 3.6×10^8J (*b*) 36×10^6J (*c*) 6.6×10^8J (*d*) 3.6×10^6J

8. *An electric lamp is marked 100 W, 220 V. The resistance of the filament will be*
(*a*) 22 Ω (*b*) 2200 Ω (*c*) 484 Ω (*d*) 4840 Ω

NUMERICALS FOR PRACTICE

1. Find out the number of electrons passing through a wire per second if the wire carries a current of 10 ampere. The charge on each electron is 1.6×10^{-19}C.
(*AMIE; W, 1987*) (**Ans.** *6.25 × 1019*)

2. A constantan wire of radius 2 mm and specific resistance 49×10^{-6} ohm cm is to be used for making a resistance of 1 ohm. Calculate the length of wire required. What will be the length of wire if the radius be doubled? (**Ans.** *25.65 m; 102.6 m*)

3. A length of copper wire of mass 5.1 kg has a resistance of 2.679 ohm. Calculate the length and diameter of wire. Density of copper is 8.93 g cm^{-3} and specific resistance of copper is 1.7 $\times 10^{-6}$ ohm - cm. (**Ans.** *300 m; 0.1557 cm*)

4. A potential difference of 100 V exists between the ends of a wire of length 200 cm. If its diameter is 0.15 mm and the current through it is 2A, calculate the specific resistance of the wire material. (*Ans. 44.175 × 10^{-6} ohm-m*)

5. There are two wires *A* and *B* of the same material. *A* is 20 times longer than *B* and has one-third of the cross-section as that of *B*. If the resistance of *A* is one ohm, what is the resistance of *B*? (*Ans. 1/60 ohm*)

6. A copper wire 150 m long has a diameter 1/3rd of the diameter of a length of manganin wire. The resistivity of the manganin and copper are 0.42 and 0.017 *micro - ohm - m* respectively. Calculate the length of the manganin wire if it has the same resistance as the copper wire.
(**Ans.** *54.645 m*)

7. Determine the resistance of a metal tube in terms of external diameter D, the internal diameter d, the length l and resistivity ρ. Hence calculate the resistance of aluminium tube 0.5 cm thick and 2 metre long. The external diameter is 15 cm and specific resistance is 3 micro ohm - cm. (**Ans.** *$4\rho l / [\pi(D^2 - d^2)]$; 26.34 $\mu\Omega$*)

8. An aluminium and a copper wire are connected together in parallel. The current flowing in the respective wires is in the ratio of 30:40. The aluminium wire is 45% longer than the copper wire and the ratio of their specific resistance is 30: 17. Find the ratio of their cross – sectional areas. (**Ans.** *1.919:1*)

9. Five cubic centimetre of copper (*i*) are drawn into a wire 150 m long, (*ii*) rolled into a square sheet of 0.1 m side. Find the resistance of the wire and the resistance between opposite faces of the plate if the specific resistance of copper is 1.72 μ ohm - cm. (**Ans.** *77.4 ohm; 8.6×10^{-10} ohm*)

10. A silver wire has a resistance of 1.28 ohm at 0°C and a temperature co-efficient of resistance of 0.00375 per degree centigrade. To what temperature must the wire be raised to double the resistance? (**Ans.** *260°C*)

11. The following are the details of the load on a circuit connected through a supply meter:
    (*i*) Six lights of 60 W each working for 5 hour per day
    (*ii*) Four fluorescent tubes of 40 W each working for 4 hour per day
    (*iii*) Two heaters of 1000 W each working for 1 hour per day.
    (*iv*) One electric iron of 750 W working for 2 hour per day.
    (*v*) Six fans of 60 W each working for 18 hour per day.

    If each unit of energy costs Rs. 2.80, what will be the total bill for the month of June. (*Ans. Rs. 1043.28*)

VIVA-VOCE/REASONING QUESTIONS

1. What is electric current?
2. Electric current is constituted by drifting of electrons from negative to positive terminal whereas the direction of flow of current is always marked from positive terminal to negative terminal, why so?
3. Out of 100 W lamp and 60 W lamp which will have smaller resistance, explain why?
4. Which will consume more energy in a given time a 40 W filament lamp or a 40 W fluorescent tube ?
5. Due to excess or deficit of electrons on a body from its normal due share, the body attains negative or positive charge on it respectively. Now the question is that we always say excess or deficit of electrons we never say excess or deficit of protons, why?

SHORT ANSWER TYPE QUESTIONS

1. What is electricity?
2. What is a charged body?

3. What do you understand by free electrons?
4. Define electric potential and mention its *SI* units.
5. What do you mean by electric resistance? Mention the factors on which it depends.
6. What is conductance? What are its units?
7. What is Ohm's law? Give its limitations.
8. How temperature affects the resistance?
9. Show that $R_1 = R_0 (1 + \alpha_0 t)$
10. What is the effect of temperature on the resistivity of copper?
11. What are the electrical, mechanical and heat units of energy and how these are related to each other?

TEST QUESTIONS

1. What is electricity?
2. Distinguish between e.m.f. and potential difference.
3. What do you understand by electric current? Give its unit. Explain why the flow of electron in an electric circuit is opposite to the direction of conventional current?
4. What are the factors affecting the resistance of a conductor? How they affect its value.
5. Write short note on resistance.
7. State and explain Ohm's law. Give its limitations.
8. What do you understand by temperature co-efficient of resistance? Specify its units. Name five materials whose resistance decreases with rise in temperature.
9. When a 100 W, 230 V filament lamp is switched *ON* to rated supply, will it draw the same current at the start and under operating condition? Justify your answer.
11. What do you understand by electrical energy and electrical power? Give their practical units.
13. Establish relations between:

    (*i*) Horse power and kW; (*ii*) kWh and kilo calories; (*iii*) Horse power and torque.

ANSWERS TO OBJECTIVE TYPE QUESTIONS

(i) Fill in the blanks:

1. inversely	**2.** decreases	**3.**less	**4.** electric energy
5. 1 kWh	**6.** more	**7.**25	**8.** decreases
9. temperature	**10.** increases	**11.** 1.6×10^{-19}	**12.** 628×10^{16}
13. 0.519.	**14.** electric energy	**15.** 6	

(ii) Multi-Choice Questions:

1. (*a*)	**2.** (*c*)	**3.** (*b*)	**4.** (*d*)	**5.** (*a*)
6. (*c*)	**7.** (*d*)	**8.** (*c*)		

CHAPTER

2 DC CIRCUITS

TOPICS TO BE DISCUSSED

Ohm's law, series-parallel resistance circuits, calculation of equivalent resistance, Kirchhoff's laws and their applications.

LEARNING OBJECTIVES

*After the completion of this **chapter,** the students/readers will be able to understand:*

1. What is Ohm's law and its applications?
2. DC Circuits and its elements.
3. Effective value of resistance, when resistors are connected in series, parallel or series-parallel combination.
4. What are the applications of series, parallel and series-parallel circuits.
5. Kirchhoff's laws.
6. How Kirchhoff's laws are used to solve electric netwarks?

INTRODUCTION

The arrangement by which various electrical energy sources, resistance and other parameters are connected together is called *electrical circuit* or *electrical network*. Various laws and theorems have been developed to analyze these simple and complex electrical circuits.

However, in this chapter, we shall discuss only Ohm's law and Kirchhoff's laws used to solve simple electric networks.

2.1 OHM'S LAW

Ohm's law *states that the current flowing between any two points of a conductor (or circuit) is directly proportional to the potential difference across them, provided physical conditions i.e. temperature etc. do not change.*

Mathematically; $I \propto V$

or $\frac{V}{I} = constant$ or $\frac{V_1}{I_1} = \frac{V_2}{I_2} = ... = \frac{V_n}{I_n} = \text{constant}$

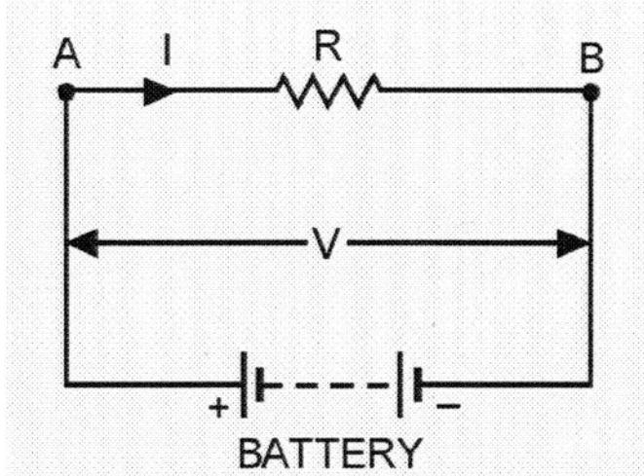

Fig. 2.1

In other words, Ohm's law can also be stated as:

The ratio of potential difference across any two points of a conductor to the current flowing between them is always constant, provided the physical conditions *i.e.* temperature etc. do not change.

This constant is called the resistance (R) of the conductor (or circuit)

$$\therefore \qquad \frac{V}{I} = R$$

It can also be written as $V = IR$; $I = \frac{V}{R}$;

In a circuit, when current flows through a resistor, the p.d. across the resistor is known as *voltage drop* across it *i.e.* $V = IR$.

Limitations of Ohm's law: Ohm's law cannot be applied to the circuits:

(*i*) Consisting of electronic tubes or transistors because these elements are not bilateral.

(*ii*) Consisting of non-linear elements such as electric are etc.

Example 2.1. *The specific resistance of platinum at 0°C is 10.5 micro - ohm - cm. What should be the length of platinum wire of No. 32 S.W.G. (diameter = 0.0274 cm) to have a resistance of 4 ohms at 0°C ?*

Solution: Resistance of a wire at 0°C;

$$R_0 = \rho_0 \frac{l}{a} \qquad \text{or} \qquad l = \frac{R_0 a}{\rho_0}$$

where, $R_0 = 4\ \Omega$

$\rho_0 = 10.5 \times 10^{-6}\ \Omega\ \text{cm} = 10.5 \times 10^{-8}\ \Omega\ \text{m}$

$$a = \frac{\pi}{4} d^2 = \frac{\pi}{4}(0.0274)^2\ \text{cm}^2 = \frac{\pi}{4}(0.0274) \times 10^{-4}\ \text{m}^4$$

$$\therefore \qquad l = \frac{4 \times \pi \times (0 \cdot 0274)^2 \times 10^{-4}}{4 \times 10 \cdot 5 \times 10^{-8}}$$

$$= \textbf{2.249 m } (\textit{Ans.})$$

2.2 D.C CIRCUITS

The closed path in which direct current flows is called **d.c. circuit.**

A simple *d.c.* circuit is shown in Fig. 2.2 which contains a *d.c.* source (battery), a load (lamp), a swtich, connecting leads and measuring instruments like ammeter and voltmeter.

The simplified line diagram of the same *d.c.* circuit is shown in Fig. 2.3. The load resistors are connected in series, parallel or series – parallel combination as per the requirement.

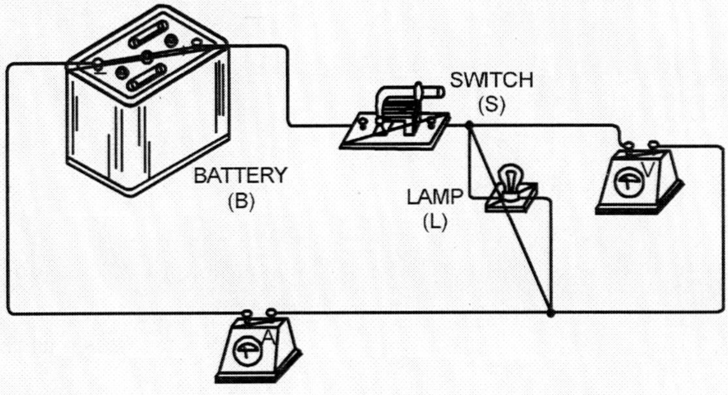

Fig. 2.2

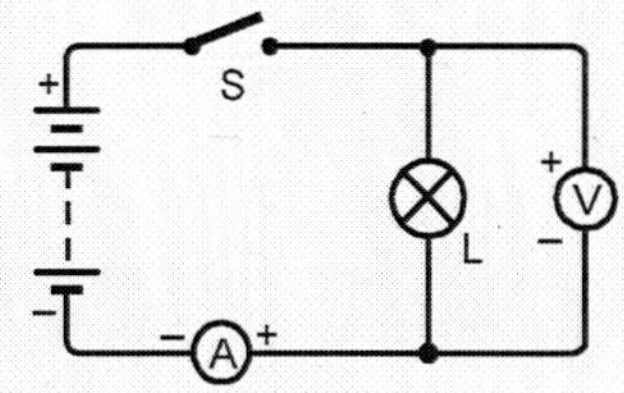

Fig. 2.3

2.3 SERIES CIRCUITS

The circuit is which number of resistors are connected end to end so that same current flows through them is called *series circuit*.

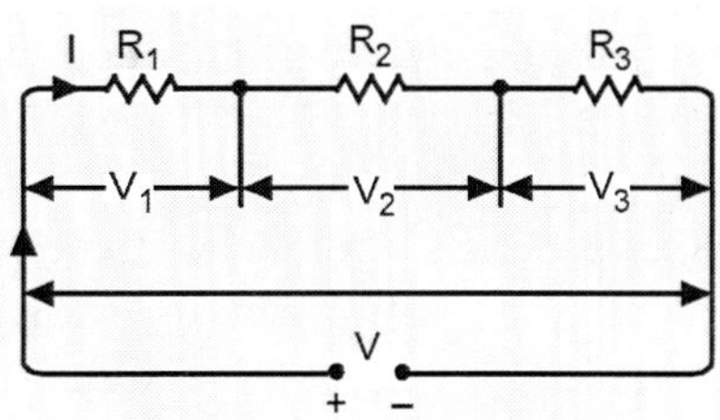

Fig. 2.4

Fig. 2.4 shows a simple series circuit. In the circuit, three resistors R_1, R_2 and R_3 are connected in series across a supply voltage of V volt. The same current (I) is flowing through all the three resistors.

If V_1, V_2 and V_3 are the voltage drops across the three resistors R_1, R_2 and R_3 respectively, then

$$V = V_1 + V_2 + V_3 = IR_1 + IR_2 + IR_3 \text{ (As per Ohm's law)}$$

Let R be the total resistance of the circuit, then

$$IR = IR_1 + IR_2 + IR_3 \quad \text{or} \quad R = R_1 + R_2 + R_3$$

i.e. Total resistance = Sum of the individual resistances.

The common application of this circuit is in the marriages for decoration purposes where a number of low voltage lamps are connected in series. In this circuit, all the lamps are controlled by a single switch, they can not be controlled individually. In domestic, commercial and industrial wiring system, the main-switch and fuses are connected in series to provide the necessary control and protection.

2.4 PARALLEL CIRCUITS

The circuit is which one end of all the resistors is joined to a common point and the other ends are also joined to another common point so that different current flows through them is called *parallel circuit*.

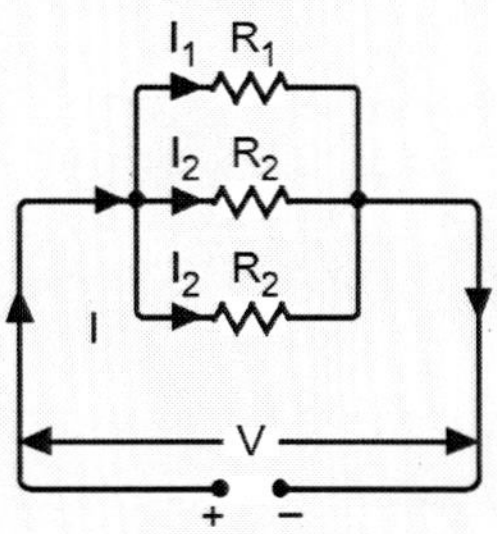

Fig. 2.5

Fig. 2.5 shows a simple parallel circuit. In this circuit, three resistors R_1, R_2 and R_3 are connected in parallel across a supply voltage of V volt. The current flowing through them is I_1, I_2 and I_3 respectively.

The total current drawn by the circuit,

$$I = I_1 + I_2 + I_3 = \frac{V}{R_1} + \frac{V}{R_2} + \frac{V}{R_3} \text{ (according to ohm's law)}$$

Let R be the total or effective resistance of the circuit, then

$$\frac{V}{R} = \frac{V}{R_1} + \frac{V}{R_2} + \frac{V}{R_3} \quad \text{or} \quad \frac{1}{R} = \frac{1}{R_1} + \frac{1}{R_2} + \frac{1}{R_3}$$

i.e. Reciprocal of total resistance = sum of reciprocal of the individual resistances.

All the appliances are operated at the same voltage, therefore, all of them are connected in parallel. Each one of them can be controlled individually with the help of a separate switch.

2.5 SERIES – PARALLEL CIRCUITS

The circuit in which series and parallel circuits are connected in series is called *series - parallel circuit*.

Fig. 2.6 shows a simple series - parallel circuit. In this circuit, two resistors R_1 and R_2 are connected in parallel with each other across terminals AB. The other three resistors R_3, R_4 and R_5 are connected in parallel with each other across terminal BC. The two groups of resistors R_{AB} and R_{BC} are connected in series with each other across the supply voltage of V volt.

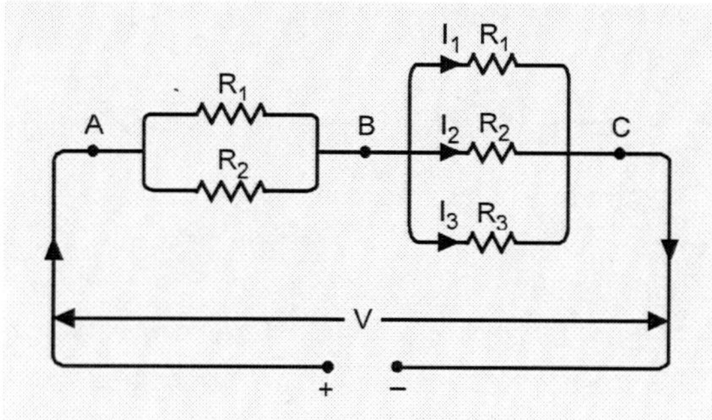

Fig. 2.6

The total or effective resistance of the whole circuit can be determined as given below:

$$\frac{1}{R_{AB}} = \frac{1}{R_1} + \frac{1}{R_2} = \frac{R_1 + R_2}{R_1 R_2} \quad \text{or} \quad R_{AB} = \frac{R_1 R_2}{R_1 + R_2}$$

Similarly,

$$\frac{1}{R_{BC}} = \frac{1}{R_3} + \frac{1}{R_4} + \frac{1}{R_5} = \frac{R_3 R_4 + R_4 R_5 + R_5 R_3}{R_3 R_4 R_5}$$

or

$$R_{BC} = \frac{R_3 R_4 R_5}{R_3 R_4 + R_4 R_5 + R_5 R_3}$$

Total or effective resistance of the circuit, $R = R_{AB} + R_{BC}$

2.6 DIVISION OF CURRENT IN PARALLEL CIRCUITS

When two resistors are connected in parallel

Fig. 2.7 shows two resistors having resistance R_1 and R_2 connected in parallel across supply voltage of V volt. Let the current in each branch be I_1 and I_2 respectively.

According to ohm's law; $I_1 R_1 = I_2 R_2 = V$ or $\frac{I_1}{I_2} = \frac{R_2}{R_1}$

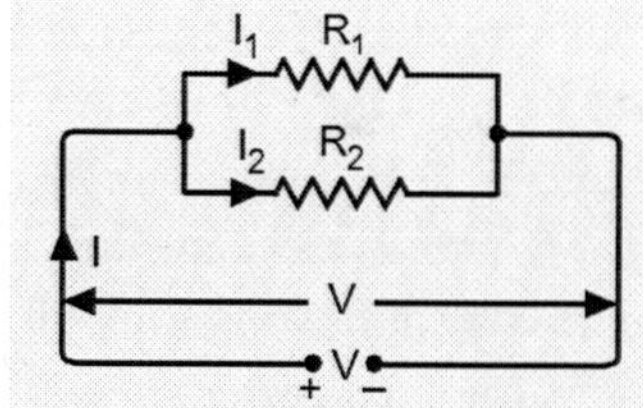

Fig. 2.7

Hence, the current in each branch of a parallel circuit is inversely proportional to its resistance respectively. The value of branch current can also be expressed (as given below) in terms of total circuit current, *i.e.*

$$I_1 R_1 = I_2 R_2 = IR = V$$

where R is total or effective resistance of the circuit and I is the total current.

$$R = \frac{R_1 R_2}{R_1 + R_2}$$

Now, $$I_1R_1 = IR = I\frac{R_1 R_2}{R_1 + R_2} \text{ or } I_1 = I\frac{R_2}{R_1 + R_2}$$

Similarly, $$I_2 = I\frac{R_1}{R_1 + R_2}$$

Example 2.2. *A resistor R is connected in series with a parallel circuit comprising of two resistors having resistance of 12 and 8 ohm respectively. The total power dissipated in the circuit is 96 watts applied voltage is 24V. Calculate the value of R.*

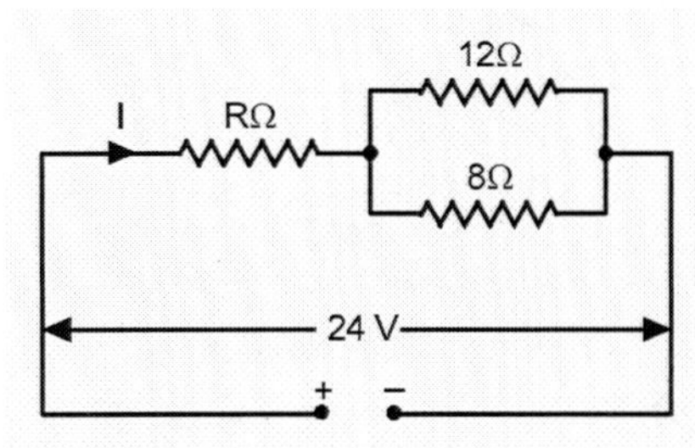

Fig. 2.8

Solution: Total power dissipated P = 96 W; Applied voltage, V = 24 V

The circuit diagram is shown in Fig. 2.8. Equivalent resistance of the two resistances connected in parallel is say R_p.

$$\therefore \quad \frac{1}{R_P} = \frac{1}{12} + \frac{1}{8} = \frac{2+3}{24} = \frac{5}{24}$$

or $$R_P = \frac{24}{5} = 4.8 \text{ ohm}$$

Current supplied to the circuit, $$I = \frac{P}{V} = \frac{96}{24} = 4 \text{ A}$$

Effective resistnace of the circuit, $$R_{eff} = \frac{V}{I} = \frac{24}{4} = 6 \text{ ohm}$$

Now $$R_{eff} = R + R_p$$

$$\therefore \quad R = R_{eff} - R_p = 6 - 4.8 = \mathbf{1.2\ ohm}\ (Ans.)$$

Example 2.3. *A circuit consists of three resistances of 12 ohm, 18 ohm and 36 ohm respectively joined in parallel is connected in series with a fourth resistance. The whole circuit is supplied at 60 V and it is found that power dissipated in 12 ohm resistance is 36 W. Determine the value of fourth resistance and the total power dissipated in the group.*

Solution: The circuit diagram is shown in Fig. 2.9 power dissipated in 12 ohm resistor, P_1 = 36 W

If the current in this resistor is I_1 ampere.

Then $\quad I_1^2 \times 12 = P_1$

$$I_1^2 = \frac{36}{12} = 3 \quad \text{or} \quad I_1 = \sqrt{3} = 1.732 \text{ A}$$

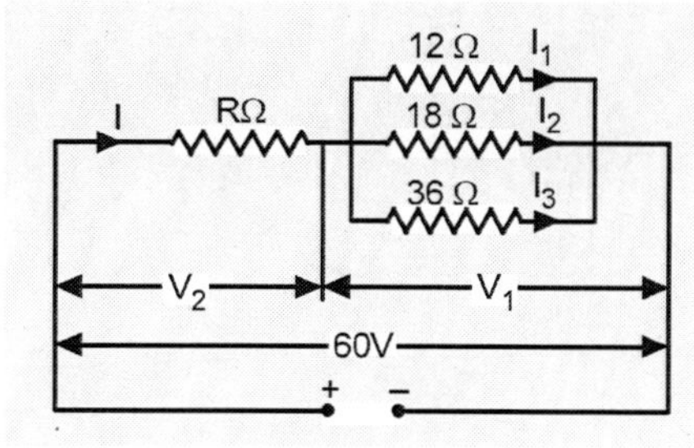

Fig. 2.9

Voltage across parallel resistors, $V_1 = I_1 \times 12$
$= 1.732 \times 12$
$= 20.785$ V

Current in 18 ohm resistor,

$$I_2 = \frac{V_1}{18} = \frac{20 \cdot 785}{18} = 1 \cdot 155 \text{ A}$$

Current in 36 ohm resistor, $\quad I_3 = \dfrac{V_1}{36} = \dfrac{20 \cdot 785}{36} = 0 \cdot 577$ A

Current in resistor $\quad R, I = I_1 + I_2 + I_3 = 1.732 + 1.155 + 0.577 = 3.464$ A

Voltage across resistor R, $V_2 = 60 - V_1 = 60 - 20.785 = 39.215$ V

$\therefore$ Value of series resistor, $\quad R = \dfrac{V_2}{I} = \dfrac{39 \cdot 215}{3 \cdot 464} =$ **11.32 ohm** (*Ans.*)

Example 2.4. *Determine current I in the circuit shown in Fig. 2.10, all the resistors are given in ohms.*

Solution: To solve this type of circuit, start from the far end of the supply. A simplified circuit is shown in Fig. 2.11.

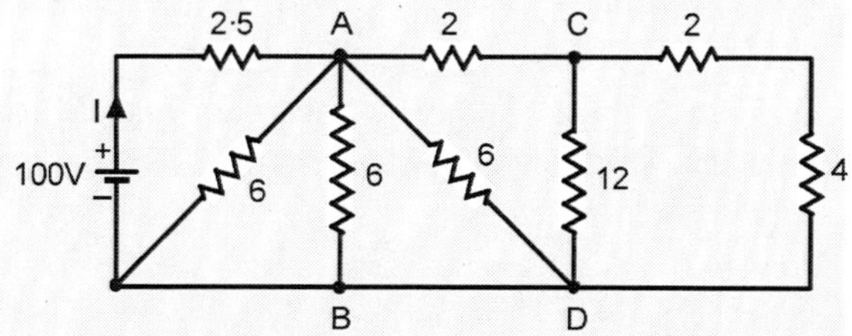

Fig. 2.10

The far end resistors of value 2Ω and 4Ω are connected in series with each other. Let their effective value be R_1 ohms.

$\therefore \quad R_1 = 2 + 4 = 6\Omega$

Then 12 Ω resistor and R_1 are connected in parallel with each other (see Fig. 2.12). Let their effective value be R_{CD}.

$$\therefore \quad R_{CD} = \frac{6 \times 12}{6 + 12} = 4\Omega$$

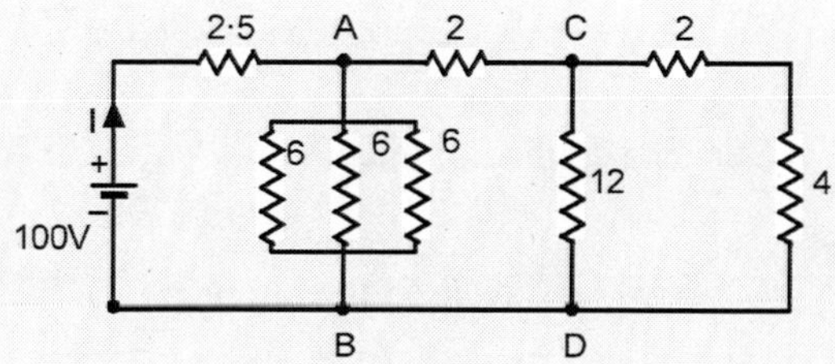

Fig. 2.11

This resistance (R_{CD}) is connected in series with 2Ω resistor as shown in Fig. 2.13. Their effecitve value is say R_2.

$\therefore \quad R_2 = 2 + 4 = 6\ \Omega$

This resistance (R_2) is connected in parallel with the three resistors of 6Ω each already connected in parallel as shown in Fig. 2.14. Their effective value is say R_{AB}.

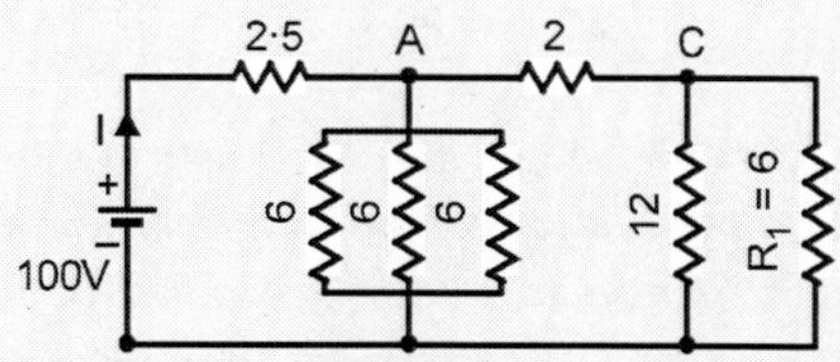

Fig. 2.12

$$R_{AB} = \frac{6}{4} = 1\cdot5\,\Omega$$

This resistance (R_{AB}) is connected in series with 2.5 Ω resistor as shown in Fig. 2.15. The total resistance of the circuit is say R ohm.

Then $$R = 2.5 + 1.5 = 4\Omega$$

∴ $$\text{Current, } I = \frac{V}{R} = \frac{100}{4} = 25\text{ A } (Ans.)$$

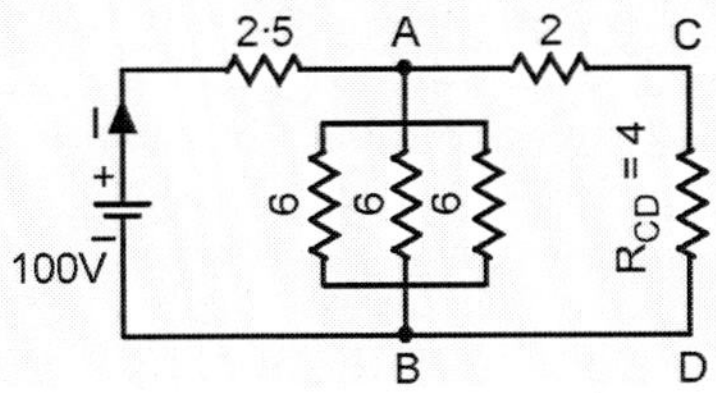

Fig. 2.13

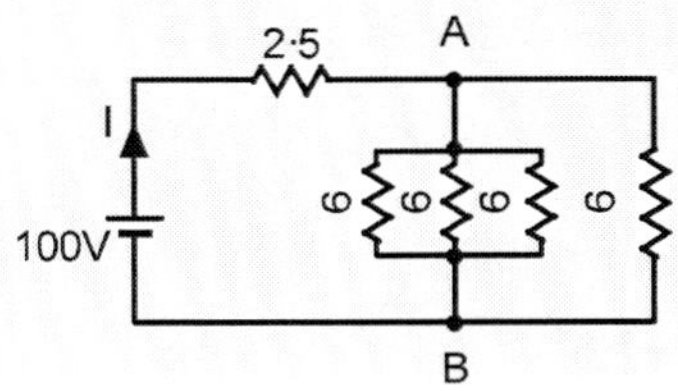

Fig. 2.14

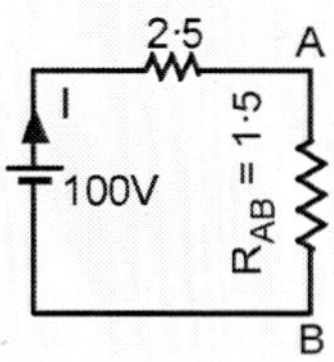

Fig. 2.15

Alternatively: After analysing the circuit, the effective resistance across the supply.

$$R = [[\{(2+4) \parallel 12\} + 2] \parallel 6 \parallel 6] + 2\cdot5$$

$$= [\{(6 \parallel 4) + 2\} \parallel 6 \parallel 6 \parallel 6] + 2\cdot5 = \left[\left(\frac{6\times12}{6+12} + 2\right) \parallel 6 \parallel 6 \parallel 6\right] + 2\cdot5$$

$$= [(4+2) \parallel 6 \parallel 6 \parallel 6] + 2\cdot5 = \left(\frac{6}{4}\right) + 2\cdot5 = 1.5 + 2.5 = 4\Omega$$

$$\text{Current, } I = \frac{V}{R} = \frac{100}{4} = 25\text{ A } (Ans.)$$

TEST YOUR LEARNING-1

Type-I: Short Answer Questions

1. When number of resistors are connected in series, show that their effective value increases.
2. When number of resistors are connected in parallel, show that their effective value decreases.
3. Give the applications of (*i*) series circuits (*ii*) parallel circuits.

Type-II: Numerical for Practice

1. An oven takes 16 A at 220 V. It is desired to reduce the current to 12 A. Find: (i) The resistance which must be connected in series. (ii) Voltage across the resistor. (*Ans.* 4.58 ohm, 54.96 V)
2. A 100W, 250V bulb is put in series with a 40W, 250V bulb across 500V supply. What will be the current drawn, what will be the power consumed by each bulb and will such a combination work? (*Ans.* 0.2286A, 32.65W, 81.63 W)

3. Two coils are connected in parallel and a voltage of 200V is applied to the terminals. The total current taken is 25A and the power dissipated in one of the coils is 1500 watts. What is the resistance of each coil. (*Ans.* 26.67 Ω, 11.43 Ω)
4. A cooking range has two coils of 500W, 250V, they can be run one at a time, both in series, both in parallel. They are run in this way for 1 hour each i.e. total 4 hours a day. Cost of 1 unit of energy being Rs. 2.50. Find the electric charges for a month of 31 days. (*Ans.* Rs. 174.38)
5. A bulb rated 110V, 60W is connected in series with another bulb rated 110 V, 100 W across 220 V mains. Calculate the resistance which should be joined in parallel with the first bulb so that both the bulbs may take their rated power. (*Ans.* 302.45 ohm)
6. A 100 ampere current is shared by 3 resistors connected in parallel. The resistor wires are of the same material and have their length in the ratio 2:3:4 and their cross-sectional area in the ratio 1: 2: 3, determine the current in each resistor. (*Ans.* 26.087 A, 34.782 A, 39.131 A)
7. An aluminium wire and a copper wire are connected in parallel. Their specific resistance are in the ratio of 30:17. The former carries 70% current more than the later and the later is 50% longer than the former. Determine the ratio of their cross-sectional area. (*Ans.* 2: 1)

Type-III: Test Questions

1. In an electric circuit, the given number of resistors may be connected in series, parallel or series-parallel combination, what is the significance of each connections?
2. When three resistors R_1, R_2, and R_3 are connected in parallel, show that the current flowing through one branch R_1 is given as, $I_1 = \dfrac{R_2 R_3}{R_1 R_2 + R_2 R_3 + R_3 R_1} \times I$.

2.7 NETWORK TERMINOLOGY

A simple electric network is shown in fig. 2.16. It contains two voltage sources E_1 and E_2 and three resistors R_1, R_2 and R_3. In fact,

The interconnection of either passive elements or the interconnection of active and passive elements constitute an **electric network.**

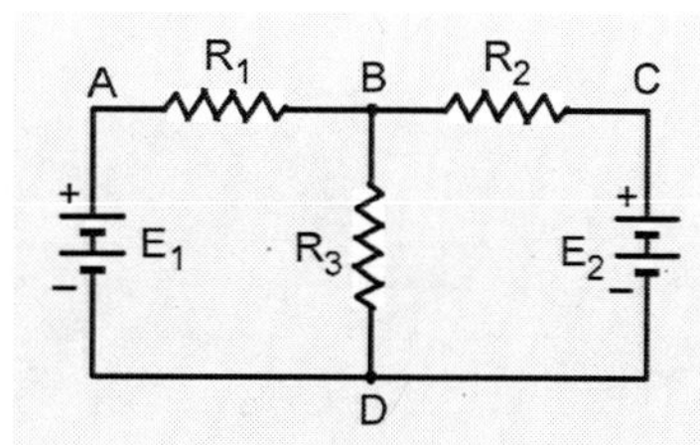

Fig. 2.16

2.7.1 Active elements

The elements which supply energy in an electric network are called *active elements*. In the circuit shown in fig. 2.16, E_1 and E_2 are the active elements.

Note: When a battery is delivering current from its positive terminal, it is under discharging condition, but if it is receiving current at its positive terminal then it is under charging condition. In both the cases, it will be considered as an active element.

2.7.2 Passive elements

The elements which receive electrical energy and dispose the same in their own way of disposal are called passive elements. In the circuit shown in fig. 2.16, R_1, R_1 and R_3 are the passive elements.

The other passive elements which are not used in this circuit are inductors and capacitors.

2.7.3 Network Terminology

Network theorems are applied to analyze the electrical network. While discussing these theorems one comes across various terms discussed below:

1. **Electric Network:** A combination of various electric elements, connected in any manner, is called an *electric network.*
2. **Electric circuit:** An *electric circuit* is a closed conducting path through which an electric current either flows or is intended to flow.
3. **Junction:** A junction is a point is the network where three or more circuit elements are joined. In fact, it is a point where current is divided. In fig.2.16, *B* and *D* are junctions.
4. **Branch:** The part of a network which lies between two junction points is called branch. In fig. 2.16. *D A B*, *B C D* and *B D* are the three branches.

2.8 KICHHOFF'S LAWS

Gustav Kirchhoff, a German scientist, summed up his findings in a set of two laws known as *Kirchhoff's laws.*

Kirchhoff's First Law: Since this law relates the currents following through the circuit, it is also known as *Kirchhoff's Current Law* (***KCL***). This law states that

The algebraic sum of all the currents meeting at a point or junction is zero.

Mathematically; $\Sigma I = 0$

To determine algebraic sum in fig 2.17, consider

Incoming currents as +ve; and Outgoing currents as –ve.

Applying Kirchhoff's current law to junction O in fig. 2.17, we get,

$$I_1 - I_2 - I_3 - I_4 + I_5 = 0$$

or $$I_1 + I_5 = I_2 + I_3 + I_4$$

i.e. Sum of incoming currents = sum of outgoing currents.

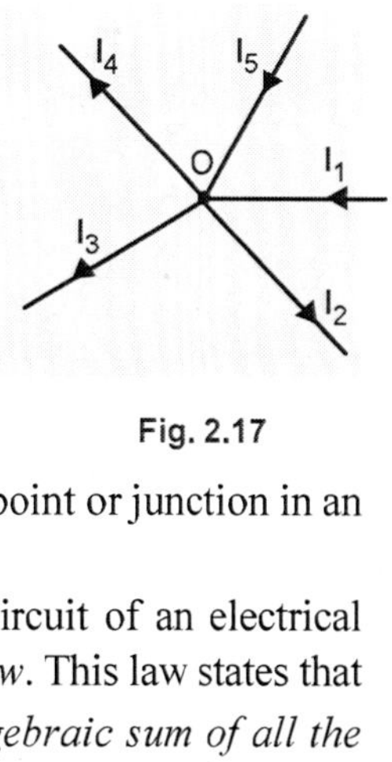

Fig. 2.17

Hence, Kirchhoff's Current Law can also be stated as:

The sum of incoming currents are equal to the sum of outgoing currents at a point or junction in an electrical network.

Kirchhoff's Second Law: Since this law relates the voltages in a closed circuit of an electrical network, it is also known as *Kirchhoff's Voltage Law* (***KVL***) or *Kirchhoff's mesh law*. This law states that

In a closed circuit or mesh, the algebraic sum of all the e.m.fs. plus the algebraic sum of all the voltage drops (*i.e.* product of current and resistances) *is zero.*

i.e. algebraic sum of all the e.m.fs. + algebraic sum of all the voltage drops = 0

Mathematically, $\sum E + \sum V = 0$ (*algebraic values*)

As algebraic sum of e.m.fs. and voltage drops to be taken, we have to determine their signs while tracing a circuit. For this, we always consider.

A rise in potential as +ve;

and a fall in potential as – ve.

Consider a branch *A B* containing only one source of e.m.f. (*E*) as shown in fig. 2.18(*a*). While tracing this branch from A to B, there is fall in potential, therefore, E is negative. However, if we trace it from B to A, there is rise in potential and E will be positive. In short,

Tracing branch A to B, E is negative (*i.e.* $-E$);

Tracing branch B to A, E is positive (*i.e.* $+E$).

Note: *Direction of flow of current is not considered since it has no effect.*

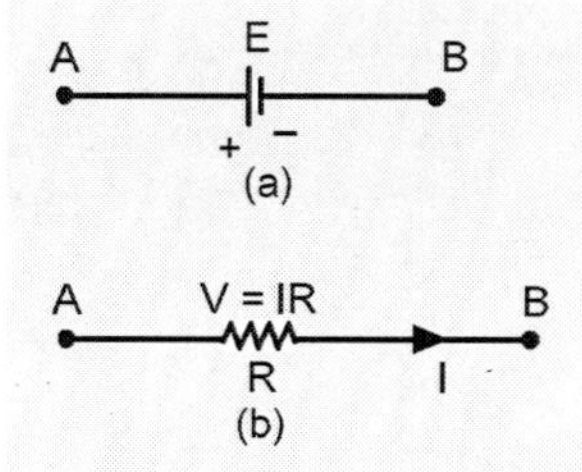

Fig. 2.18

To determine sign for the voltage drop V ($= IR$), consider a branch containing resistor of resistance R in which current I flows from A to B as shown in fig. 2.18(b). Therefore, A is at higher potential w.r.t. terminal B.

Thus,

Tracing branch from A to B, V is negative (*i.e.* $-V$) as there is *fall in potential.* Since we are moving in the direction of flow of current.

Tracing branch from B to A, V is positive (*i.e.* $+V$) as then is *rise in potential.* Since we are moving opposite to the direction of flow of current.

Note: *Only the direction of flow of current determine the sign of V.*

Illustration:

Consider a network shown in fig. 2.19. In this circuit, we can apply KVL for the closed circuit $BCDA$ and form an equation given below:

$$-I_1 R_1 - E_2 + I_2 R_3 = 0$$

Where $I_1 R_1$ (voltage drop) is considered as negative, since we are tracing the circuit along the direction of flow of current.

E_2 (e.m.f. of battery) in taken as negative as we move from +ve to –ve terminal at the battery and there in fall in potential.

$I_2 R_3$ (voltage drop) in taken as positive as we move opposite to the flow of current.

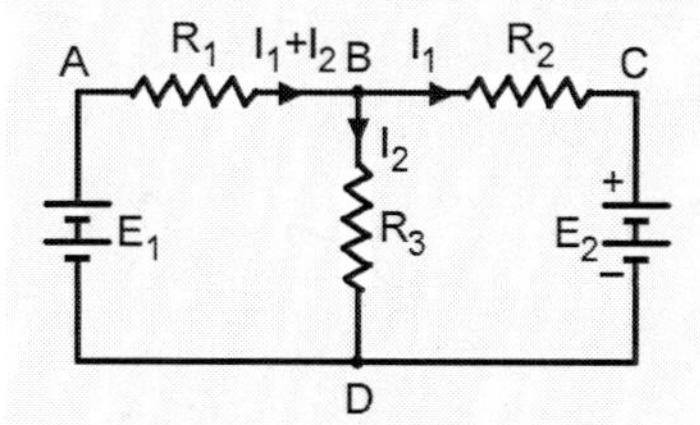

Fig. 2.19

Similarly, applying (kVL) to the closed circuit $ABDA$, we get, $-(I_1 + I_2) R_1 - I_2 R_3 + E_1 = 0$

SOLUTION OF NETWORK BY KIRCHHOFF'S LAWS

When a network is to be solved by applying Kirchhoff's laws, the following steps are taken:

1. Mark the assumed direction of flow of current in various branches of the network according to KCL.
2. Choose as many number of closed circuits, as the number of unknown quantities.
3. Applying KVL to the chosen closed circuits and prepare the equations.
4. After solving the equations, determine the unknown values.

Note: *If the determined current carries the – ve, sign, it shows that the actual direction of flow of current is opposite to that of the assumed direction of flow of current in the given branch.*

2.9 WHEATSTONE BRIDGE

For the first time, Wheatstone (an English telegraph engineer) proposed this bridge for measuring the value of an unknown resistance. This bridge consists of four arms AB, BC, AD and DC having resistances

P, Q, X and R respectively (see fig. 2.20). Resistance P and Q are the known (fixed value) resistances and called *ratio arms*. Whereas, the resistance R is a variable resistance of known value and X is an unknown resistance whose value is to be determined.

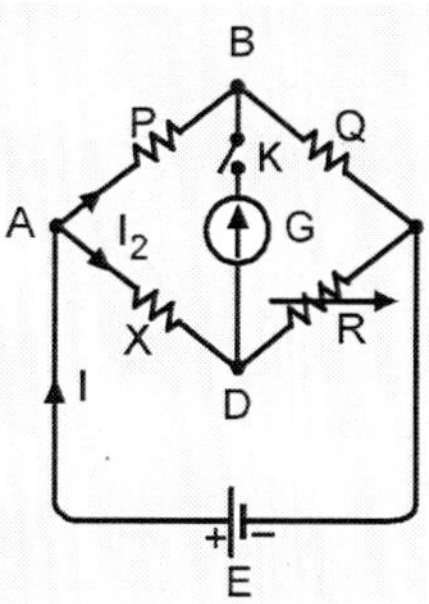

Fig. 2.20

To determined the value of X, connect a battery E across A and C a galvanometer G across B and D through key K. The bridge is said to be balanced when galvanometer G gives zero defection on closing the key K. The balance is obtained by selecting the values of resistors P, Q suitably and finally adjusting the value of R.

At balance;

Curent flowing through galvanometer = 0

Current flowing through resistor P and $Q = I_1$

Current flowing through resistor X and $R = I_2$

Voltage drop across AB = Voltage drop across AD

$$I_1P = I_2X \quad ...(i)$$

Voltage drip across BC = Voltage drop across DC

$$I_1Q = I_2R \quad ...(ii)$$

Dividing equation (*i*) by (*ii*), we get,

$$\frac{P}{Q} = \frac{X}{R} \quad \text{or} \quad XQ = PR$$

i.e. Product of opposite arms = Product of other opposite arms

or
$$X = \frac{P}{Q} \times R$$

Example 2.5. *In the circuit shown in fig. 2.21. calculate current and power in the 6 ohm resistor.*

Solution: The various junction points are marked in fig. 2.22. According to Kirchhoff's first law, let the current flowing through various sections be a shown in fig. 2.22.

Applying Kirchhoff's second law to mesh *ABEFA*, get,

$$-20(I_1 + I_2) - 6I_2 + 140 = 0 \quad \text{or} \quad 20I_1 + 26I_2 = 140$$

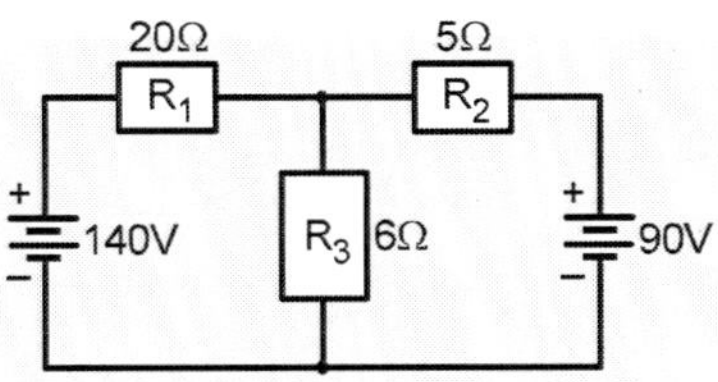

Fig. 2.21

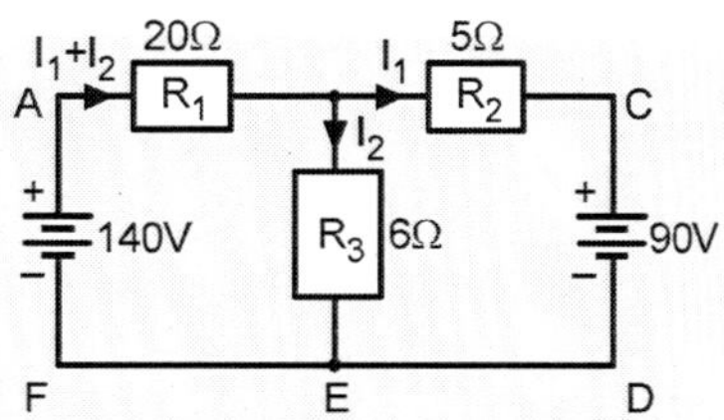

Fig. 2.22

or
$$10I_1 + 13I_2 = 70 \quad ...(i)$$

Applying Kirchhoff's second law to mesh *BCDEB*, we get,

$$-5I_1 - 90 + 6I_2 = 0$$

or $$-5I_1 + 6I_2 = 90$$

or $$-10I_1 + 12I_2 = 180 \qquad ...(ii)$$

Adding equation (*i*), and (*ii*), we get

$$25I_2 = 250$$

$$I_2 = \mathbf{10\ A}\ (Ans.)$$

Power in 6 ohm resistor $= I_2^2 R_3 = (10)^2 \times 6 =$ **600 W** (*Ans.*)

Example 2.6. *Fig.2.23 shows two batteries connected in parallel each represented by an emf along with its internal resistance. A load resistance of* 6 Ω *is connected across the ends of the batteries. Calculate the current through each battery and the load.*

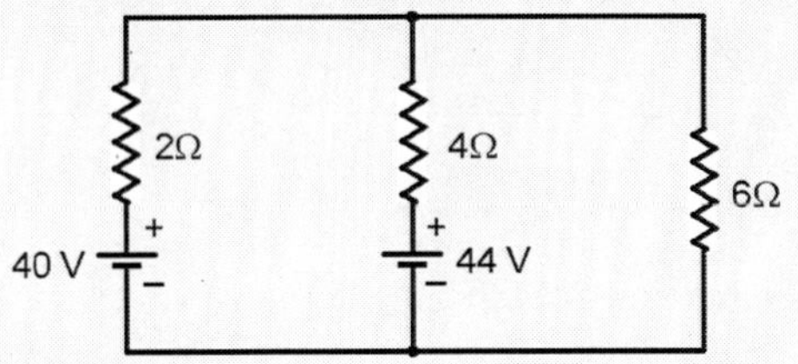

Fig. 2.23

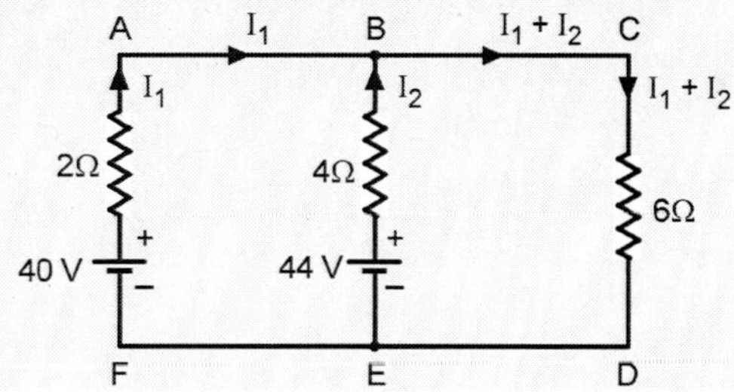

Fig. 2.24

Solution: Let the current flowing through various branches be as marked in fig. 2.24.

Applying *KVL* to mesh *A B E F A*, we get

$$-2I_1 + 4I_2 - 44 + 40 = 0$$

or $$2I_1 - 4I_2 = 40 - 44 \quad \text{or} \quad 2I_2 - I_1 = 2 \qquad ...(i)$$

Applying *KVL* to the mesh *BCDEB*, we get,

$$-4I_2 - 6(I_1 + I_2) + 44 = 0$$

or $$4I_2 + 6(I_1 + I_2) = 44 \quad \text{or} \quad 5I_2 + 3I_1 = 22 \qquad ...(ii)$$

Multiplying equation (*i*) by 3 and adding with equation (*ii*) we get,

$$I_2 = \frac{28}{11}\text{A}\ (Ans.)$$

Substituting the value of I_2 in equation (*i*),

$$I_1 = \frac{34}{11}\text{A}\ (Ans.);$$

$$\text{Current through load, } I_L = I_1 + I_2 = \frac{34}{11} + \frac{28}{11} = \frac{62}{11}\text{ A} = \mathbf{5.636\ A}\ (Ans.)$$

Example 2.7. *Two batteries A and B are connected in parallel and a load of* 10 Ω *is connected across their terminals. A has an emf of* 12 *V and an internal resistance of* 2 Ω *and B has an emf of* 8 *V and an internal resistance of* 1 Ω. *Use Kirchhoff's laws to determine the magnitude of currents and also the directions in each of the batteries. Also determine the pd across external resistance.*

Solution: Let the current flowing through various branches be as marked in fig. 2.25.

Applying Kirchhoff's voltage law to mesh *ABCDA*, we get,

$$2I_1 - I_2 + 8 - 12 = 0$$

or $\quad 2I_1 - I_2 = 12 - 8$

or $\quad 2I_1 - I_2 = 4 \qquad ...(i)$

Applying Kirchhoff's voltage law to mesh *ADCEA*, we get,

$$I_2 + 10(I_1 + I_2) - 8 = 0$$

or $\quad 10I_1 + 11I_2 = 8 \qquad ...(ii)$

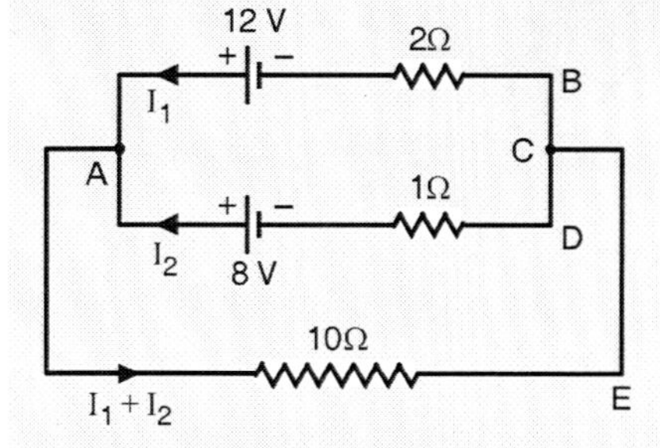

Fig. 2.25

Solving equations (*i*) and (*ii*), we get,

$$I_1 = 1.625 \text{ A } (Ans.)$$

and $\quad I_2 = -0.75 \text{ A } (Ans.)$

The negative sign with current I_2 shows that the current I_2 is flowing opposite to the assumed direction.

Current flowing through the load resistance of 10 Ω = $I_1 + I_2 = 1.625 + (-0.75) = 0.875$ A

PD across load resistance $= (I_1 + I_2)R = 0.875 \times 10 = 8.75$ V (*Ans.*)

Example 2.8. *Solve the network shown in fig. 2.26 for the current through 6 ohm resistor.*

Solution: Let the current flowing through various branches be as marked in fig. 2.27.

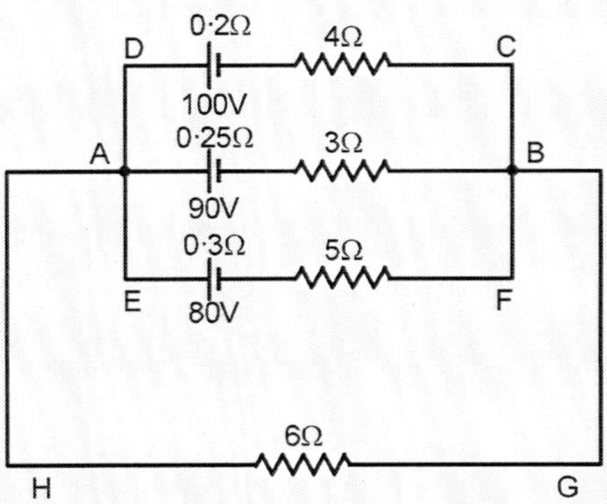

Fig. 2.26

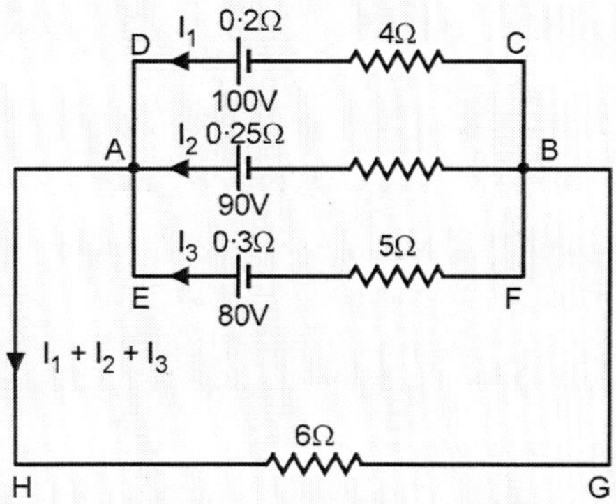

Fig. 2.27

Applying Kirchhoff's second law to the following closed circuits,

Circuit CDAHGBC

$$-4I_1 - 0.2I_1 + 100 - 6(I_1 + I_2 + I_3) = 0$$

or $\quad 10.2I_1 + 6I_2 + 6I_3 = 100 \qquad ...(i)$

Circuit BAHGB

$$-3I_2 - 0.25I_2 + 90 - 6(I_1 + I_2 + I_3) = 0$$

or $\quad 6I_1 + 9.25I_2 + 6I_3 = 90 \qquad ...(ii)$

Circuit *FEAHGBF*

$$-5I_3 - 0.3I_3 + 80 - 6(I_1 + I_2 + I_3) = 0$$

or $\quad 6I_1 + 6I_2 + 11.3I_3 = 80 \qquad ...(iii)$

Subtracting (*ii*) from (*i*), we get,

$$4.2I_1 - 3.25I_2 = 10 \qquad ...(iv)$$

Solving (*ii*) and (*iii*), we get,

$$31.8I_1 + 68.525I_2 = 537 \quad ...(v)$$

From equation (*iv*) and (*v*), we get,

$$391.51I_2 = 137.4$$

$$\therefore \quad I_2 = \frac{1937 \cdot 4}{391 \cdot 15} = 4.953 \text{ A}$$

Solving equation (*v*), we get, $I_1 = 6.21$ A

Solving equation (*iii*), we get, $I_3 = 1.15$ A

Current in 6 ohm resistor, $I = I_1 + I_2 + I_3$

$= 6.21 + 4.953 + 1.15 =$ **12.313 A** (*Ans.*)

Example 2.9. *A Wheatstone bridge consists of AB = 4 ohm, BC = 3 ohm, CD = 6 ohm and DA = 5 ohm. A 2 volt cell is connected between B and D and a galvanometer of 10 ohm resistance between A and C. Find the current through the galvanometer.*

Solution: The circuit is shown in fig. 2.28.

Applying Kirchhoff's first law at junction *B*, *A* & *C*, the current in various branches is marked.

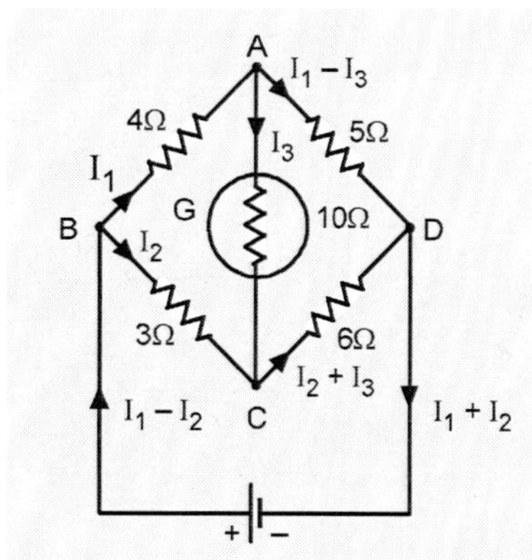

Fig. 2.28

Applying Kirchhoff's second law to various closed loops;

Considering loop *BACB*, we get,

$$-4I_1 - 10I_3 + 3I_2 = 0$$

or $$4I_1 - 3I_2 + 10I_3 = 0 \quad ...(i)$$

Considering loop *ADCA*, we get,

$$-5(I_1 - I_3) + 6(I_2 + I_3) + 10I_3 = 0$$

or $$-5I_1 + 5I_3 + 6I_2 + 6I_3 + 10I_3 = 0$$

or $$5I_1 - 6I_2 - 21I_3 = 0 \quad ...(ii)$$

Considering loop *BADEB*, we get

$$-4I_1 - 5(I_1 - I_3) + 2 = 0$$

or $$-4I_1 - 5I_1 + 5I_3 = -2$$

or $$9I_1 - 5I_3 = 2 \quad ...(iii)$$

Multiplying equation (*i*) by 2 and subtracting from equation (*ii*), we get,

$$5I_1 - 6I_2 - 21I_3 = 0$$

$$8I_1 - 6I_2 + 20\,I_3 = 0$$

$$- \quad + \quad -$$

$$-3I_1 \qquad -41I_3 = 0$$

$$\therefore \quad I_1 = -\frac{41}{3} I_3$$

Substituting the value of I_1 in equation (*iii*), we get,

$$9\left(-\frac{41}{3} I_3\right) - 5I_3 = 2 \quad \text{or} \quad -123I_3 - 5I_3 = 2 \quad \text{or} \quad I_3 = \frac{-1}{64} \text{ A}$$

Current flowing through galvanometer is 1/64 ampere from C to A.

TEST YOUR LEARNING-2

Type-I: Short Answer Questions

1. What do you mean by electric network?
2. What do you understand by active and passive elements of an electric network?
3. State Kirchhoff's law.
4. State Kirchhoff's second law.

Type-II: Numerical for Practice

1. Two batteries *A* and *B* connected in parallel and load of 10Ω is connected across their terminals. A has an e.m.f. of 12 V and an internal resistance of 2Ω; *B* has an e.m.f. of 8V and an internal resistance of 1 Ω. Using Kirchhoff's laws determine the value and direction of flow of current in each battery and in the external resistance. Also determine the potential difference across the external resistance. *(Ans. 1.625 A charging; 0.75 A discharging; 0.875 A; 8.75 V)*
2. A Wheatstone bridge *ABCD* is arranged as follows:
   Resistance between $A-B$, $B-C$, $C-D$, $D-A$ and $B-D$ are 10, 2, 8, 4 and 5 ohm respectively. A 100 V supply is connected between terminals *A* and *C*. Determine the currents in branches *AB* and *AD* of the circuit and total current taken from the supply. *(Ans. 7.44 A; 11.905 A; 19.345 A)*
3. Find the current in all the branches of the circuit shown in fig. Q.3. *(Ans. 4.4915 A, 2.2881 A, 2.2084 A)*
4. For the circuit shown in fig. Q.4, find the current in 7.5Ω resistor. *(Ans. 0.662 A)*

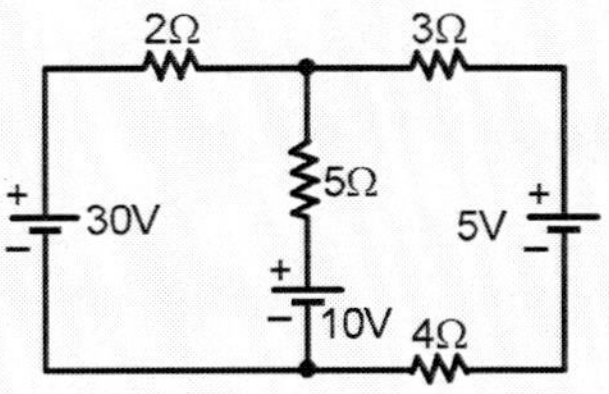

Fig. Q.3

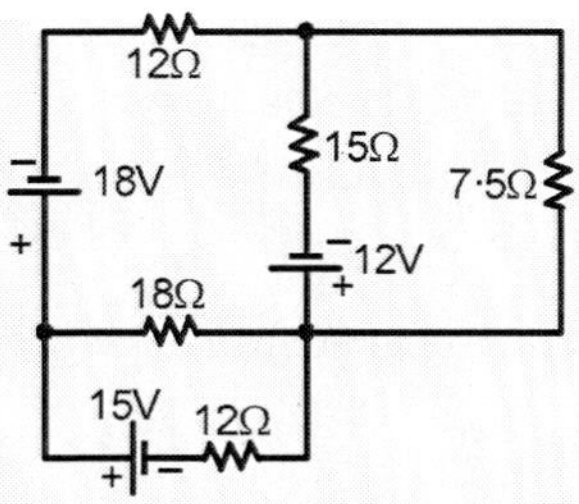

Fig. Q.4

Type-III: Test Questions

1. Draw and explain an electric-network. Draw a network and show a node, junction and a branch.
2. Define and explain Kirchhoff's laws of electric circuit.

SUMMARY

1. *Ohm's law*: Ratio of V to I is constant for a circuit provided physical conditions remain the same $R = \frac{V}{I}$; $V = IR$ or $I = \frac{V}{R}$
2. *Effective value of the resistance* (R)
   (*i*) in series circuit; $R = R_1 + R_2 + ... + R_n$
   (*ii*) in parallel circuit; $\frac{1}{R} = \frac{1}{R_1} + \frac{1}{R_2} + ... + \frac{1}{R_n}$
3. *Kirchhoff's current law (KCL):* It states that the algebraic sum of all the currents meeting at a junction is always zero.
4. *Kirchhoff's voltage law (KVL)*: It states that the algebraic sum of all the emfs and voltage drops in a closed circuit is always zero.

SELF ASSESMENT

TEST YOUR PREPARATION

OBJECTIVE TYPE QUESTIONS

Type-I: Fill in the blanks:

1. A man has seven resistors each of value 1/7 ohm. The maximum value which he can obtain by their combination will be __________ ohm.
2. Two resistors of 10 ohm each, are connected in parallel, the combination is connected in series with another 5 ohm resistor, the effective resistance of the circuit will be __________ ohm.
3. All the domestic appliances are connected in __________ with the supply.
4. A wire has a resistance of x ohm and carries a current of y ampere, the voltage across the wire will be __________ volt.
5. According to *KCL*, at any junction of an electrical network, the sum of __________ currents = sum of out going currents.
6. The voltage drop in a resistor will be taken as __________ if we are tracing the circuit in the direction of flow of current.

Type-II: Multi-choice questions:

1. *A man has three resistances, each of value 1/3 ohm. What is the minimum resistance he can obtain by combining them?*
   (*a*) 1 ohm (*b*) 3 ohm (*c*) 1/6 ohm (*d*) 1/9 ohm

2. *If n identical resistors are connected in series, their total value is K ohm. What will be their effective value if they are now connected in parallel.*
(a) K/n^2 (b) K^2/n (c) K/n (d) n^2/K
3. *A parallel arrangement of 3 ohm and 6 ohm resistors is placed in series with an 8 ohm resistor. If p.d. of 60 V is connected across the whole circuit, the current in the three ohm resistor is*
(a) 6 A (b) 4 A (c) 2 A (d) 10 A
4. *Four wires of equal length and of resistance 20 ohm each are connected in the form of a square. The equivalent resistance between the two opposite corners of the square is*
(a) 10 Ω (b) 20 Ω (c) 40 Ω (d) 20/4 Ω
5. *Given three equal resistors. How many different combinations of those three can be made*
(a) two (b) four (c) six (d) eight
6. *Two electric lamps each of 40 W, 230 V are connected in series across 230 V supply. The power consumed by each*
(a) 20 W (b) 10 W (c) 40 W (d) 80 W
7. *Three 3 ohm resistors are connected in the form of an equilateral triangle. The total resistance between any two corners is*
(a) 2 ohm (b) 6 ohm (c) 3 ohm (d) 4/3 ohm
8. *Three electric lamps of 60 W each are connected in parallel. The power consumed by the combination will be*
(a) 30 W (b) 120 W (c) 20 W (d) 180 W
9. *Kirchhoff's first law states that at a junction in an electric network*
(a) $\Sigma E = 0$ (b) $\Sigma I = 0$ (c) $\Sigma V = 0$ (d) $\Sigma E + \Sigma V = 0$
10. *KVL states that in a closed circuit of an electric network*
(a) $\Sigma E = 0$ (b) $\Sigma I = 0$ (c) $\Sigma V = 0$ (d) $\Sigma E + \Sigma V = 0$
11. *The polarity of the voltage drop across a resistor is determine by*
(a) value of resistor (b) value of current
(c) direction of current (d) all *a*, *b*, and *c*

NUMERICALS FOR PRACTICE

1. Calculate the resistance between the points *A* and *B* in network shown in fig. Q.1. All the resistors are given in ohms. (*Ans. 2 ohm*)

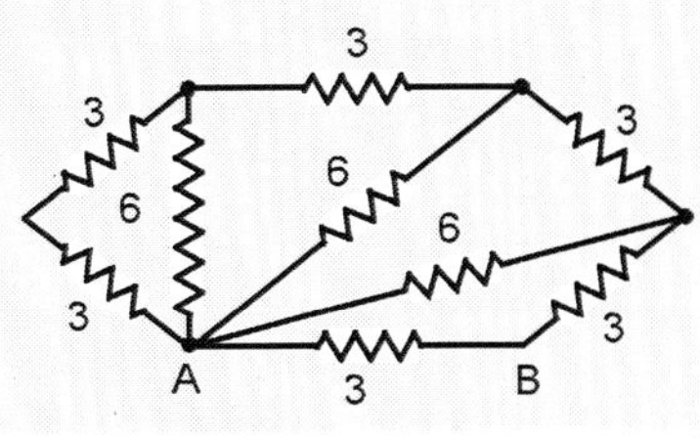

Fig. Q.1

2. Each resistance of the network shown in fig. Q.2 is equal to *R*. What is the resistance between the terminal *A* and *B*?
Consider that the p.d. between *C* and *E* is zero. (*Ans. R ohm*)
3. Obtain the total power supplied by the 60 V source and the power absorbed in each resistor in the network shown in fig. Q.3.
(*Ans. 360W, 252W, 27W, 54W, 15.75 W, 11.25W*)

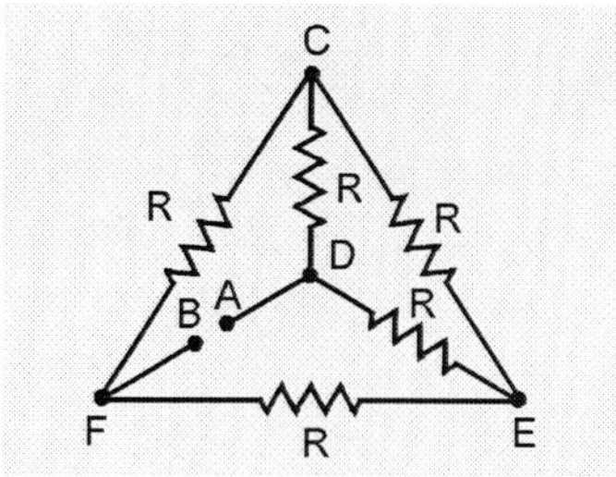

Fig. Q.2

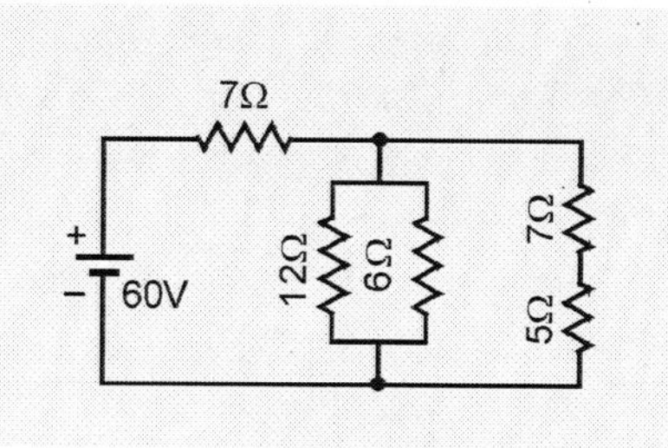

Fig. Q.3

4. A resistance of 10 Ω is connected in series with two resistances each of 15 Ω arranged in parallel. What resistance must be connected across the parallel combination that the total current taken shall be 1.5 A with 20 V applied. *(Ans. 6 ohm)*

5. In the network shown in fig. Q.5 determine current and voltage drop across 40 ohm resistor. *(Ans. 1A, 40V)*

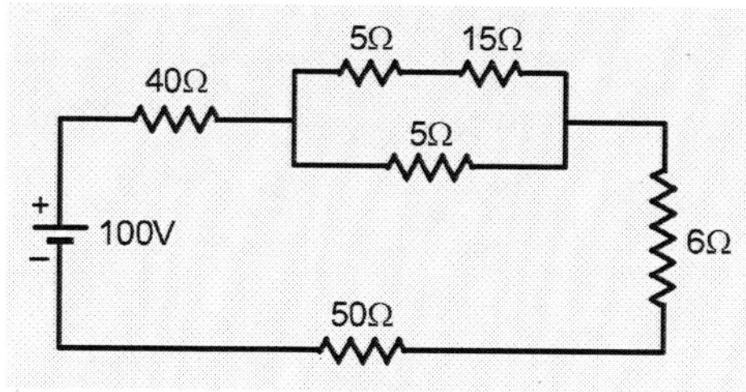

Fig. Q.5

6. A letter *A* is constructed of a uniform wire of resistance 2 ohm per cm. The sides of the letter are 60 cm long and the cross piece is 30 cm long while the apex angle is 60°. Find the resistance of the letter between the two ends of the legs. *(Ans. 150 ohm)*

7. A circuit consists of three resistances of 12Ω, 18 Ω and 36 Ω respectively joined in parallel is connected in series with a forth resistance. The whole circuit is supplied at 60V and it is found that power dissipated in 12 Ω resistance is 48 W. Determine the value of fourth resistance and the total power dissipated in the circuit. *(Ans. 9 Ω; 240 W)*

8. How many 60 W lamps may be safely run on a 230 V circuit fitted with a 5 ampere fuse. *(Ans. 19)*

9. Determine the current supplied by the battery in circuit shown in fig. Q.9. *(Ans. 29.17 mA)*

10. How many 60W lamps may be safely operated on a 230 V mains fitted with a 5A fuse? *(Ans. 19)*

11. With the help of Kirchhoff's laws, determine current in *XY* in the circuit shown in fig. Q.11. *(Ans. 40 A)*

12. Determine current supplied by the source in the network shown in fig. Q.12 using Kirchhoff' laws. *(Ans. 72.3 mA)*

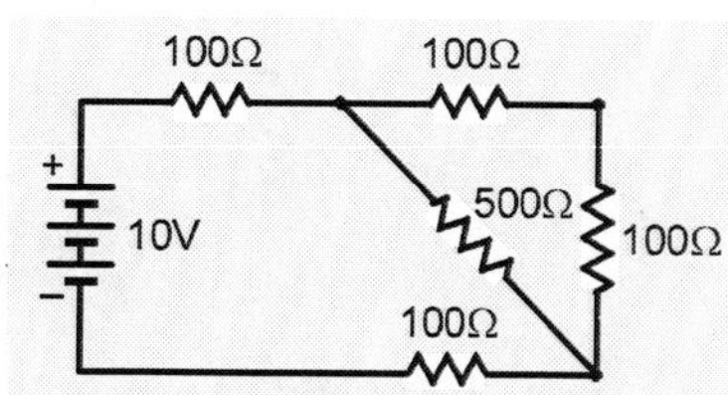

Fig. Q.9

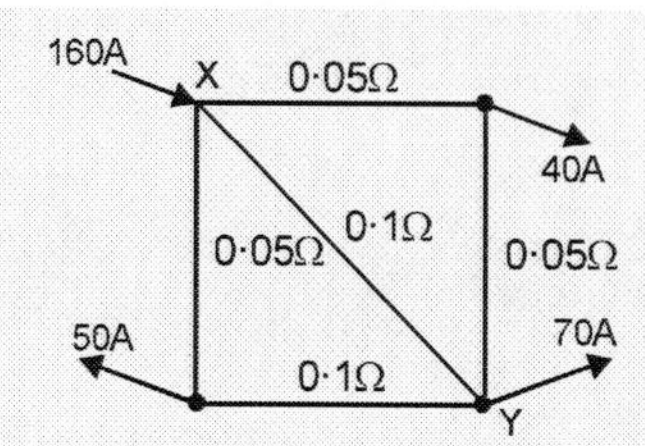

Fig. Q.11

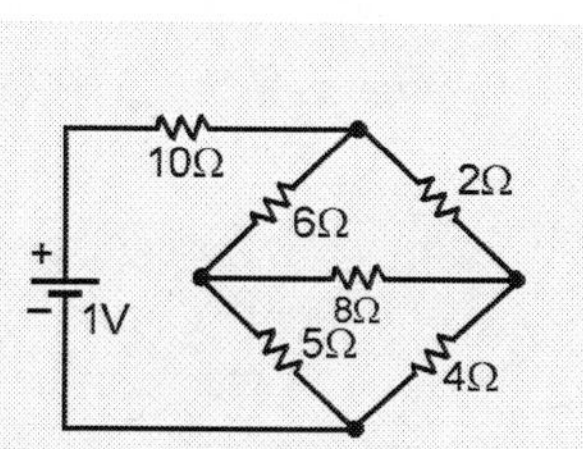

Fig. Q.12

13. *With the help of Kirchhoff's laws, determine current in* 10 *ohm resistor in the circuit shown in fig.*Q.13. *(Ans. 1.352 A)*

14. *Using Kirchhoff's laws, calculate the current in the* 6 Ω *resistor, in the circuit shown in fig.*Q.14. *(Ans.0.5 A from B to A)*

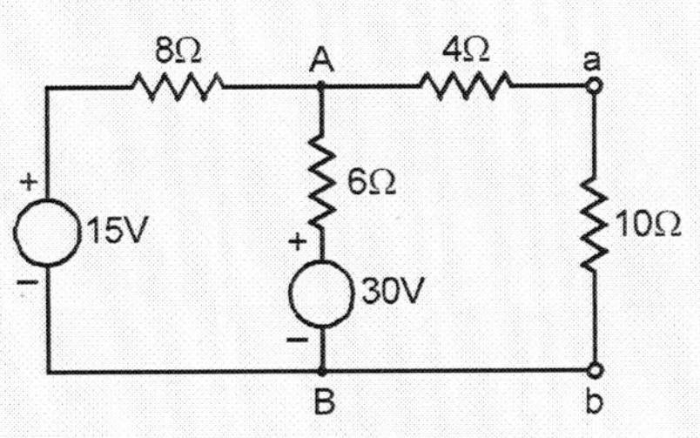

Fig. Q.13

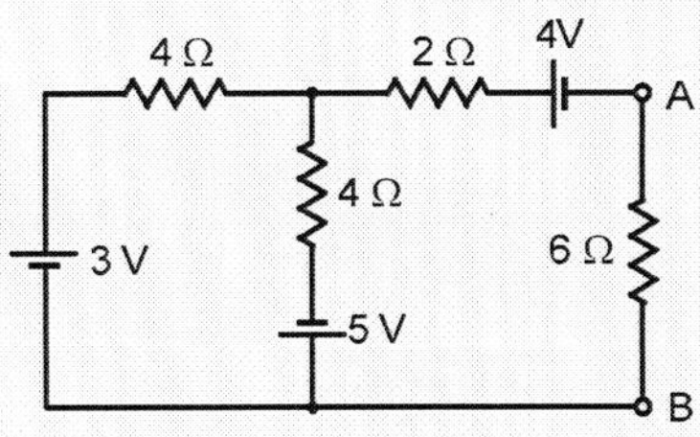

Fig. Q.14

VIVA-VOCE/REASONING QUESTIONS

1. All the electrical appliances are connected in parallel, why?
2. When load on the electric supply increases, the circuit resistance decreases, how?
3. Algebraic additions are different to numerical additions, how?

SHORT ANSWER TYPE QUESTIONS

1. What is Ohm's law? Give its limitations.
2. Show that in parallel circuits $\frac{1}{R} = \frac{1}{R_1} + \frac{1}{R_2} + \frac{1}{R_3} + \ldots\ldots + \frac{1}{R_n}$
3. State Kirchhoff's current and voltage laws.

TEST QUESTIONS

1. What are the practical applications of (*i*) series circuits and (*ii*) parallel circuits?
2. Show that:

   (*i*) in series circuit, the effective value of resistance; $R = R_1 + R_2 + \ldots\ldots\ldots + R_n$.

   (*ii*) in parallel circuit, the reciprocal of the effective value of resistance,

   $$\frac{1}{R} = \frac{1}{R_1} + \frac{1}{R_2} + \ldots\ldots\ldots + \frac{1}{R_n}$$

3. Give reasons why all the equipment are connected in parallel to the supply.
4. State and explain Kirchhoff's current and voltage law.
5. While applying Kirchhoff's voltage law to a loop, how signs are applied to the e.m.f. and voltage drop?

6. Describe the working of a Wheatstone bridge.
7. What steps are taken while solving a network by Kirchhoff's laws?

ANSWERS TO OBJECTIVE TYPE QUESTIONS

(i) Fill in the blanks:

1. one **2.** 10 **3.** parallel **4.** xy **5.** incoming
6. negative

(ii) Multi-Choice Questions:

1. (*d*) **2.** (*a*) **3.** (*c*) **4.** (*c*) **5.** (*b*)
6. (*d*) **7.** (*a*) **8.** (*d*) **9.** (*b*) **10.** (*d*)
11. (*c*)

CHAPTER

3 ELECTRIC CELLS

TOPICS TO BE DISCUSSED

Primary cell, wet cell, dry cell, battery, Li-ion battery, series and parallel connections of cells, Secondary cells, Lead Acid Cell, Discharging and recharging of cells, preparation of electrolyte, care and maintenance of secondary cells.

LEARNING OBJECTIVES

After the completion of this ***chapter,*** *the students/readers will be able to understand:*

1. What is an electric cell and how it is formed?
2. What are the primary and secondary cells?
3. What is the meaning of wet and dry cells?
4. How the cell are connected in series, parallel or in series-parallel combination and what we achieved out of these combinations.
5. What is a battery?
6. What are the constructional features of a lead-acid battery?
7. How to check the charged condition of a battery?
8. How a lead-acid battery is charged?
9. How Lithium-iron cell or battery works and what are its merits?

INTRODUCTION

Current does not flow only through metallic conductors but it also flows though some liquids called *electrolytes. Such liquids provide a large number of oppositely charged ions and the conduction is due to the movement of these ions (not due to the movement of free electrons). This conduction of current through electrolyte produces chemical changes. Thus, in this process electrical energy is converted into chemical and the converse of this is also true. The chemical effect of electric current has many applications *e.g.* extraction of pore metals from ores, electro-deposition, electroplating, electrotyping, production of oxygen, storage batteries etc.

---

* Electrolytes are the liquids which break up into oppositely charged atoms called ions e.g. acids (H_2SO_4, HCl etc.)

One of the major application of chemical effect of electric current is in storage batteries. Battery is the only device in which we can store electrical energy in the form of chemical energy.

In this chapter, we shall deal with electric cells, storage batteries and their applications.

3.1 ELECTRIC CELL

A source of e.m.f. (d.c.) in which chemical energy is converted into electrical energy is called an **electric cell.**

Forming of a cell

An electric cells essentially consists of the following:

(*i*) Two metal plates (electrodes) of different material — so that different potentials are built-up when chemical action takes place on them.

(*ii*) A suitable solution (electrolyte) such as an acid, alkali or salt solution. The solution must be capable to react chemically with the two electrodes.

When the two electrodes are immersed in a suitable electrolyte, different chemical actions take place on them and a p.d. is established between them.

E.M.F. developed in a cell

The magnitude of e.m.f. of a cell depends upon:

(*i*) nature or material of the plates used as electrodes.

(*ii*) nature or type of electrolyte used.

3.2 TYPES OF CELLS

Broody, electric cells may be divided into following two categories:

(*i*) Primary cells and

(*ii*) Secondary cells

Primary cells: *The cells in which chemical action is not reversible are called* **primary cells** *e.g.* Voltaic cell, Denial cell, Lachlanche cell, dry cell etc.

In these cells, during discharging, one of the plates (usually negative plate) is consumed which can not be recovered by reversing the direction of flow of current through the cell. Thus, chemical action in this case is not reversible and the cells cannot be recharged. This fact makes the primary cells rather an expensive source of electrical energy and rarely used in commercial applications.

3.3 WET CELLS

A *wet cell or battery* has a liquid electrolyte. It is also called as *flooded cel*l, since the liquid covers all internal parts, or *vented cell*, since gases produced during operation can escape to the air. Wet cells may be primary cells (non-rechargeable) or secondary cells (rechargeable).

3.4 DRY CELLS

A dry cell is a type of cell or battery, commonly used for portable electrical devices. In dry cells, a paste electrolyte, with only enough moisture to allow current to flow is used. Unlike a wet cell, a dry cell can operate in any orientation without spilling, as it contains no free liquid, making it suitable for portable equipment. On the other hand wet cell needs careful handling to avoid spillage. Lead–acid batteries did not achieve the safety and portability of the dry cell until the development of the gel battery.

However, wet cells are mostly used for high-drain applications, such as starting internal combustion engines, because inhibiting the electrolyte flow tends to reduce the current capability.

A common dry cell is the zinc-carbon cell, sometimes called the dry Leclanché cell, with a nominal voltage of 1.5 volt, the same as the alkaline cell (since both use the same zinc–manganese dioxide combination).

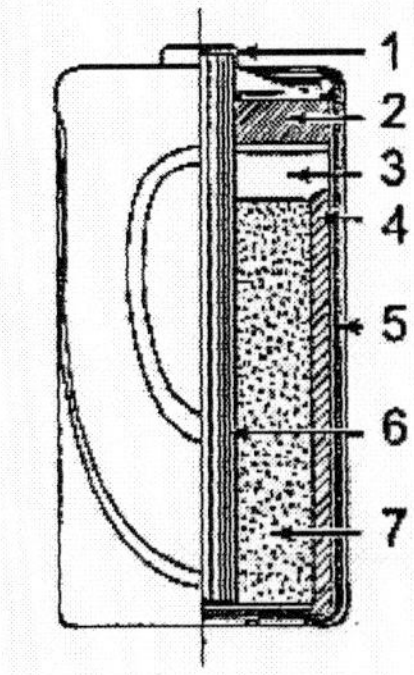

Fig. 3.1

1. Brass cap, 2. Plastic seal,
3. Expansion space, 4. Porous cardboard,
5. Zinc can, 6. Carbon rod,
7. Chemical mixture.

A standard dry cell comprises a zinc anode, usually in the form of a cylindrical pot(5), as shown in fig. 3.1, with a carbon cathode in the form of a central rod (6). The electrolyte is ammonium chloride in the form of a paste next to the zinc anode (7). The remaining space between the electrolyte and carbon cathode is taken up by a second paste consisting of ammonium chloride and manganese dioxide, the latter acting as a depolariser. In some designs, the ammonium chloride is replaced by zinc chloride.

Zinc–carbon batteries were the first commercial dry batteries, developed from the technology of the wet Leclanché cell. They made flashlights and other portable devices possible, because the battery can function in any orientation. They are still useful in low drain or intermittent use devices such as remote controls, flashlights, clocks or transistor radios. Zinc–carbon dry cells are single-use primary cells.

3.5 IMPORTANT TERMS RELATING TO AN ELECTRIC CELL

The following are the important terms related to an electric cell:

(*i*) **Electro - motive force (e.m.f.) of a cell:** The energy supplied by a cell to one coulomb of charge is called *e.m.f. of a cell*. In general, it is defined as a p.d. between the two electrodes (positive and negative) of a cell on open circuit. It is represented by letter '*E*'.

(*ii*) **Internal resistance of a cell:** The opposition offered to the flow of current by the internal composition of the cell itself is called its *internal resistance*. It is generally represented by letter '*r*'. Fig. 3.2(*a*) shows a cell with its internal resistance.

(*iii*) **Terminal voltage:** The potential difference across the terminals of a cell at load is known as its *terminal voltage*. It is generally represented by letter '*V*'.

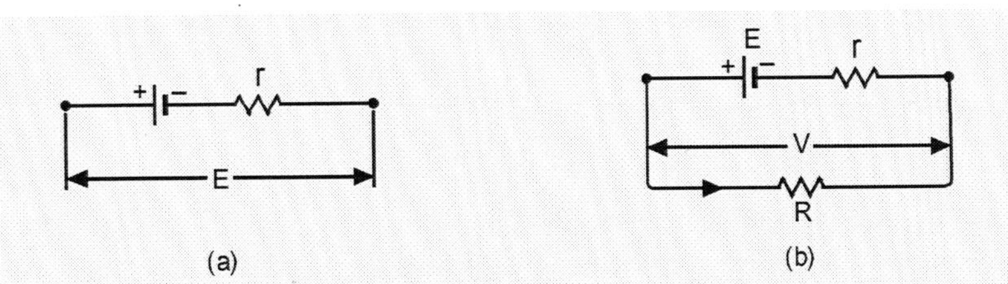

Fig. 3.2

When load is applied, it delivers current I to the external (load) resistor R as shown in fig. 3.2(*b*). The p.d. across the terminals reduces to V volts because of the voltage drop ($I \times r$) in the internal resistance of the cell.

Thus, $V = E - I \times r$

3.6 GROUPING OF CELLS

A single cell can supply a very small current at low voltage. Generally, to operate electrical and electronic circuits, a large current at high voltage is required. Therefore, a number of cells are connected in series, parallel and series – parallel grouping.

***(i)* Series grouping:** When a number of cells are connected in such a way that the negative terminal of one is connected to positive terminal of the other and so on, as shown in fig. 3.3, the cells are said to be connected in *series grouping*.

Where,

n = No. of cell connected in series;

E = *E.M.F.* of each cell;

r = Internal resistance of each cell;

R = Load resistance;

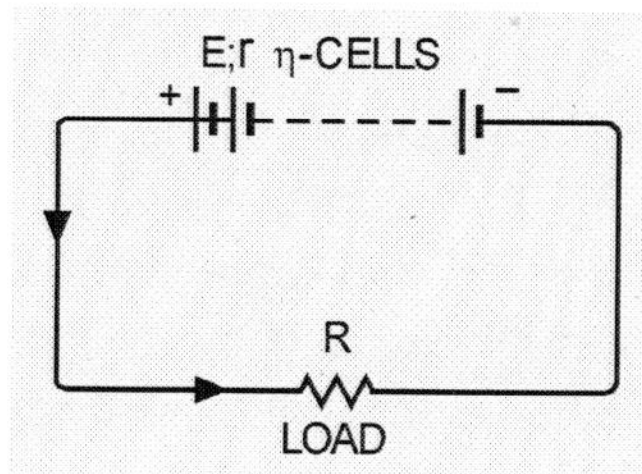

Fig. 3.3

Total e.m.f. of the grouping = nE

Total internal resistance of the grouping = nr

Total resistance of the circuit = $R + nr$

Current delivered to the load, $I = \dfrac{nE}{R + nr}$

This grouping is used where higher voltages are required.

***(ii)* Parallel grouping:** When a number of cells are connected in such a way that the positive terminals of all the cell are connected together and negative terminals are connected together separately, as shown in fig. 3.4, the cells are said to be connected in *parallel grouping*.

where, m = No. of cells connected in parallel;

E = *E.M.F.* of each cell;

r = Internal resistance of each cell;

R = Load resistance.

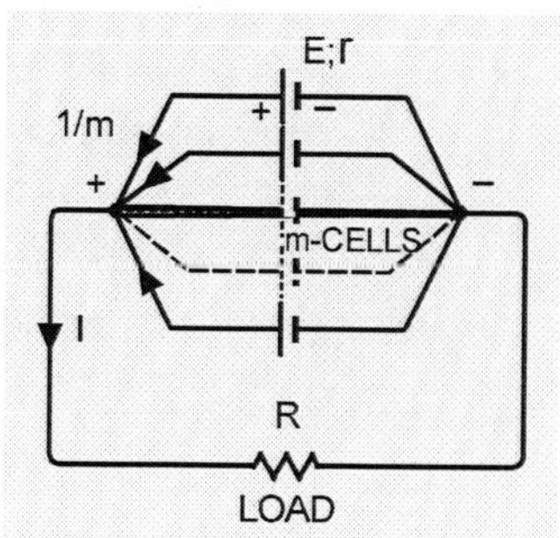

Fig. 3.4

Total e.m.f. of the grouping = E

Total resistance of the grouping = r/m

Total resistance of the circuit = $R + (r/m)$

Current delivered to the load, $I = \dfrac{E}{R + (r/m)}$

Current supplied by each cell = I/m

This grouping is used where higher currents are required.

***(iii)* Series - parallel grouping:** The grouping in which a number of cells are connected in series in one row and a number of such rows are connected in parallel is called a *series-parallel grouping* of cells.

Fig. 3.5 shows a series-parallel grouping of cells.

Where, n = No. of cells connected in series in each row;

E = *E.M.F.* of each cell;

r = Internal resistance of each cell;

m = No. of rows connected in parallel;

R = Load resistance.

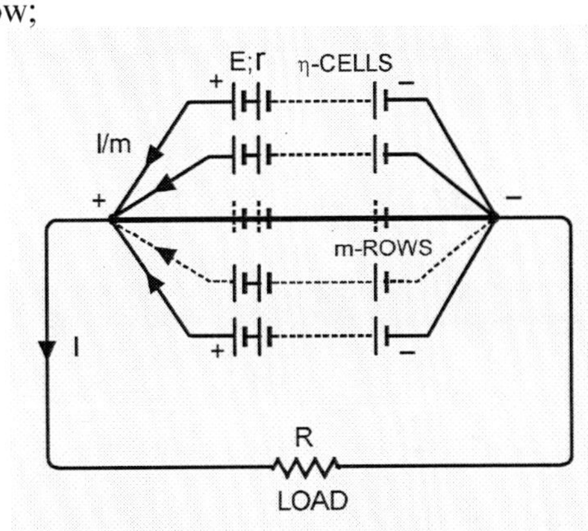

Fig. 3.5

Total e.m.f. of each row or grouping = nE

Total internal resistance of each row = nr

Total internal resistance of grouping = nr/m

Total resistance of the circuit = $R + (nr/m)$

Current delivered to the load, $I = \dfrac{nE}{R+(nr/m)}$

Current supplied by each row = $\dfrac{I}{m}$

The current supplied by a series – parallel grouping will be maximum when

Load resistance = Internal resistance of the grouping

i.e. $$R = \frac{nr}{m}$$

This grouping used where both higher currents and higher voltages are required.

Example 3.1: *How many cells, each having an e.m.f. of 1·5V and internal resistance of 0·25 ohm would be required (connected in series) to pass a current of 1·5 A through a resistance of 15 ohm.*

Solution: Let n be the number of cells connected in series:

Current delivered to the load, $I = \dfrac{nE}{R+nr}$

Here, $E = 1\cdot5$V; $R = 15\Omega$; $r = 0\cdot25\ \Omega$ and $I = 1\cdot5$ A

$$\therefore \quad 1\cdot5 = \frac{n\times1\cdot5}{15+n\times0\cdot25} \quad \text{or} \quad n = \frac{22\cdot5}{1\cdot125} = 20\ (Ans.)$$

Example 3.2: *Four dry cells each of which has an e.m.f. of 1·5V and an internal resistance of 0·06 ohm are connected in parallel. Determine current and power dissipated in the external load resistance of 2·985 ohm.*

Solution: In parallel circuit;

Current delivered to load resistance, $I = \dfrac{E}{R+(r/m)}$

Here, $E = 1\cdot5$V; $R = 2\cdot985\Omega$; $r = 0\cdot06\Omega$ and $m = 4$

$$I = \frac{1\cdot5}{2\cdot985+(0\cdot06/4)} = \mathbf{0\cdot5A}\ (Ans.)$$

Power dissipated in external load resistor;

$$= I^2R = (0\cdot5)^2 \times 2\cdot985 = 0\cdot74625\text{W}\ (Ans.)$$

Example 3.3: *Four batteries are connected in parallel to supply current to a resistive load for heating. Each battery consists of twelve cells connected in series. The e.m.f. of each cell is 2 volt and internal resistance 0·15 ohm. What should be the resistance of the load so that power consumed in the load is 307·2 W.*

Solution: Current supplied by series – parallel grouping

$$I = \frac{nE}{R + nr/m}$$

Here, $n = 12$; $E = 2\ V$; $m = 5$; $r = 0{\cdot}15\Omega$

$$I = \frac{12 \times 2}{R + (0{\cdot}15 \times 12/5)} \quad \text{or} \quad I = \frac{24}{R + 0{\cdot}36}$$

Power consumed in the load resistor, $P = I^2 R$

$$307{\cdot}2 = \left(\frac{24}{R + 0{\cdot}36}\right)^2 \times R$$

or $$R^2 + 0{\cdot}72R + 0{\cdot}1296 = 1{\cdot}875\,R \quad \text{or} \quad R^2 - 1{\cdot}155\,R + 0{\cdot}1296 = 0$$

or $$R = \frac{1{\cdot}55 \pm \sqrt{1{\cdot}344 - 0{\cdot}5184}}{2} = \frac{1{\cdot}155 \pm 0{\cdot}90312}{2};$$ $R = \mathbf{1{\cdot}029\ \Omega}$ or $\mathbf{0{\cdot}12594\ \Omega}$ (*Ans.*)

Substituting the value of R in equation (*i*), we get,

$$I = 17{\cdot}28\text{ A} \quad \text{or} \quad 49{\cdot}39\text{ A}$$ (*Ans.*)

3.7 SECONDARY CELLS

The cells in which chemical action is reversible are called **secondary cells** *e.g.* lead-acid cell, nickel-iron alkaline cell (or Edison cell), nickel-cadmium-alkaline cell etc.

In these cells, no electrode is consumed during discharging however, the chemical composition of the plates is changed. When the direction of flow of current through the cell is reversed (charging), the plates and electrolyte regain their original composition. Thus, chemical action in this case is reversible which is stored in the cell itself. For this reason, these cells are called storage cells or accumulators and used in almost all the commercial applications.

3.8 BATTERY

A series, parallel or series-parallel grouping of cells is called a **battery.**

Generally, a cell can deliver a small current at low voltage. For a circuit, if higher voltage is required – a battery containing number of cells connected in series is applied; if higher current in required a battery containing number of cells connected in parallel is applied; if larger current at higher voltage is required a battery containing number of cells in series further connected in parallel is applied.

Usually, a number of cells connected in series placed in a single container is called **a battery.**

3.9 LEAD-ACID BATTERY

Fig. 3.6 shows the cut-away view of 6V commercial lead-acid battery. The following are the important parts of the battery.

1. **Container:** It is the outer body of the battery. It is made of a hard rubber or plastic material and is sealed at the top to prevent spilling of the electrolyte. A large space is left at the bottom of the container so that the sediments that drop from the plates are collected here and may not short circuit the positive and negative plates.

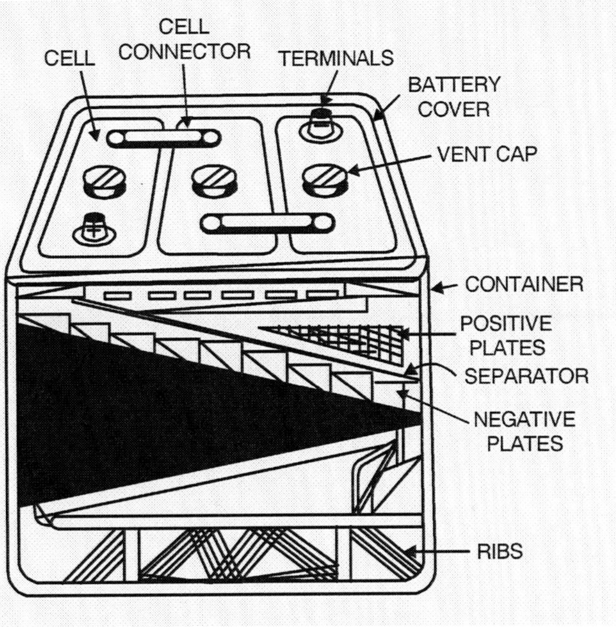

Fig. 3.6

2. **Plates:** Generally, alloy of lead-antimony sheets covered with lead-peroxide and spongy lead forming positive and negative plates respectively are used as electrodes. To increase the capacity of the battery, we use a large number of plates in each cell instead of only two plates. The number of positive* and negative plates (*i.e.* 11, 13, 15 or 17) of each cell are alternatively placed and sandwitched with an insulator called *separator* as shown in fig. 3.7(*a*). One group of positive and negative plates forms a cell which develops an e.m.f. of 2.0 volt. A separate compartment is provided for each cell in the container of the battery.

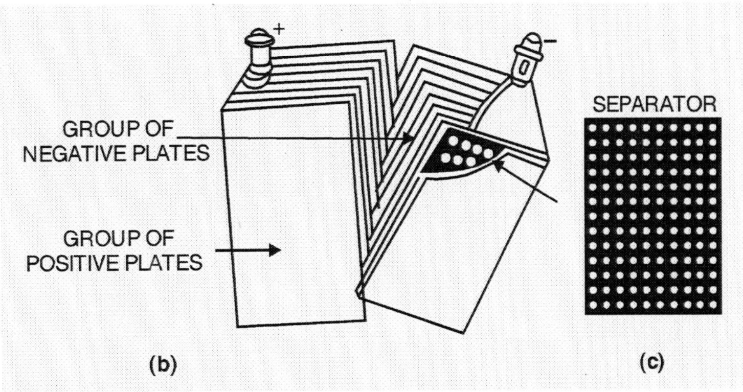

Fig. 3.7

3. **Separator:** To reduce the internal resistance of the cell and to save the space, the plates are placed very close to each other. To prevent the plates touching each other if they wrap or buckle, they are separated by a rubber sheet (non-conducting material) having large number of small holes called *separator* [see fig. 3.7(*b*)].

4. **Electrolyte:** Dilute sulphuric-acid (H_2SO_4) is used as an electrolyte in lead-acid batteries. Sulphuric-acid is added to water in such a proportion that with a fully charged battery, its specific gravity is about 1.28 to 1.29.

* The number of negative plates is always one *more* than the positive plates.

5. **Battery cover:** Each cell compartment is covered usually with a molded hard rubber and the joints between covers and container are sealed with an acid-resistance material. In each cell cover openings are provided — two for positive and negative terminals, and third for a vent. The whole container is fitted with a leak proof cover.
6. **Vent caps:** The vent-cap has a vent hole to allow free exit of the gases formed in the cell during charging. The vent caps can be easily removed for adding water. The vent cap is also removed to insert the nozzle of hydrometer for checking the specific gravity of electrolyte to check the battery charge condition.
7. **Inter-cell connector:** The cells, placed in the same container are connected in series with a lead alloy link called inter-cell connector.
8. **Cell terminals:** Each cell has two terminals which are generally made of lead as it does not corrode due to the electrolyte. The positive terminal of the battery is marked with a red colour or by a large positive (+) sign.

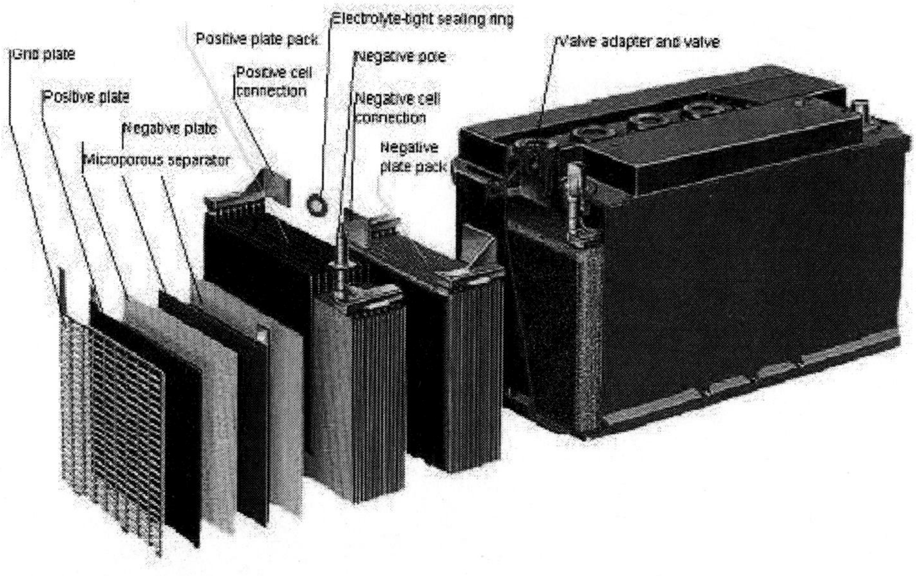

Fig. 3.8: Pictorial (cut-away) view of a car battery (Lead-acid battery)

3.10 WORKING PRINCIPLE OF LEAD-ACID CELL

When a lead-acid cell is ready for use, its positive plate is made of load peroxide (PbO_2) chocolate brown in colour and the negative plate is of spongy lead (Pb) grey in colour. Both the plates are immersed in a dilute sulphuric acid (H_2SO_4) of specific gravity about 1.28. When a load is connected across the terminals of the cell, it starts delivering current to the load, the process is called *discharging* of cell. In this process, the chemical energy stored in the cell is converted into electrical energy which is delivered to the load (external circuit).

Chemical action during discharging

The sulphuric acid when dissolved in water, its molecules are dissociated into hydrogen ions ($2H^+$) and sulphate ions (SO_4^-) which move freely (at random) in the electrolyte. When the load (resistor) is connected across the terminals as shows in Fig. 3.9 (*a*), the sulphate ions move towards the plate made of lead and hydrogen ions towards the plate made of lead peroxide. The following chemical action takes place the two electrodes.

At lead plate $SO_4^- - 2e \longrightarrow SO_4$ (radical)

The electrons travel through external load and after doing work reach at other electrode. The sulphate *radical*, chemically reacts with the electrode material (*Pb*) and forms lead sulphate ($PbSO_4$).

$$\underset{\text{(Grey in colour)}}{Pb} + SO_4 \longrightarrow \underset{\text{(whitish in colour)}}{PbSO_4}$$

At PbO_2 plate

Each hydrogen ion (H^+) on reaching the PbO_2 plate takes one electron (the same electron which is given up by the sulphate ion at *Pb* state and has come to the other plate *via* external load resistor) from it to become hydrogen gas;

$$\underset{\text{(ion)}}{2H^+} + 2e \longrightarrow \underset{\text{(gas)}}{2H}$$

the hydrogen gas liberated at this electrode acts chemically with electrode material (PbO_2) and reduces it to *PbO* which further reacts with H_2SO_4 forming $PbSO_4$.

$$\underset{\text{(Chocolate brown in colour)}}{PbO_2} + 2H \longrightarrow PbO + H_2O$$

$$PbO + H_2SO_4 \longrightarrow \underset{\text{(Whitish in colour)}}{PbSO_4} + H_2O$$

Thus during discharging;

(*i*) Both the plates are converted to lead sulphate ($PbSO_4$), whitish in colour.

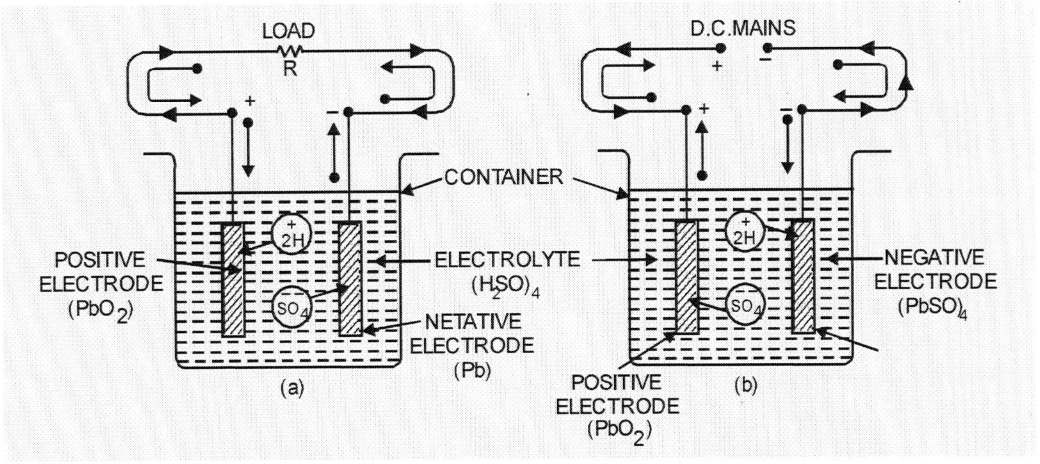

Fig. 3.9

(*ii*) Sulphuric acid is consumed and water is formed which reduces the specific gravity of H_2SO_4 from 1.28 to 1.15.

(*iii*) Terminal voltage per cell falls to 1.8V form 2.0V.

(*iv*) Chemical energy is converted into electrical energy which is delivered to the load.

Chemical action during recharging

For recharging, position plate is connected to positive terminal of the source and negative plate is connected to the negative terminal as shown in fig. 3.9 (*b*). During recharging, hydrogen ions ($2H^+$) move towards the negative plate and sulphate ions (SO_4^-) towards the positive plate. The following chemical actions take place at the two plates:

At positive plate

$$SO_4^- - 2e \longrightarrow SO_4 \text{ (radical)}$$
$$SO_4 + H_2O \longrightarrow H_2SO_4 + O$$

The oxygen (in atomic state) reacts chemically with this plate material ($PbSO_4$):

$$\underset{\text{(Whitish in colour)}}{PbSO_4} + O + H_2O \longrightarrow \underset{\text{(Chocolate brown in colour)}}{PbO_2} + H_2SO_4$$

At negative plate

Each hydrogen ion (H^+) on reaching the negative plate, takes one electron (the same electron which is given-up by the sulphate ion at positive plate and has come to negative plate *via* external circuit (*i.e.* source) from to become hydrogen gas; $2H^+ + 2e \longrightarrow 2H$

The hydrogen gas liberated at negative plate acts chemically with its material ($PbSO_4$);

$$\underset{\text{(Whitish in colours)}}{PbSO_4} + 2H \longrightarrow \underset{\text{(Grey in colour)}}{Pb} + H_2SO_4$$

Thus during recharging:

(*i*) Both the plates regain their original composition *i.e.* anode is converted back into lead peroxide (PbO_2) and cathode into spongy lead (Pb).

(*ii*) Water is consumed and sulphuric acid is formed which improves the specific gravity of H_2SO_4 from 1.15 to 1.28.

(*iii*) Terminal voltage pre cell increases from 1.8 V to 2.2/2.5 V.

(*iv*) Electrical energy is converted into chemical energy which is stored in the cell.

TEST YOUR LEARNING-1

Type-I: Short Answer Questions

1. What is an electric cell?
2. What are the requirements to form a cell?
3. How will you categories the electric cells?

4. What is the necessity of connecting the cells in series, parallel or series-parallel combination?
5. How will you differentiate an electric cell and battery?
6. Name at-least five major parts of a lead-acid battery.
7. Write down the chemical reactions occurring at positive and negative plate of a lead-acid cell during discharging.
8. Write down the chemical reactions occurring at positive and negative plate of a lead-acid cell during re-charging.

Type-II: Numerical for Practice

1. Six lead-acid cells each having an emf of 2V and internal resistance 0.05 ohm are connected in series. How much current will be delivered by the combination to a load resistance of 3.9 ohm. (*Ans.* 3A)
2. Six dry cells each of 1.5 V and an internal resistance of 0.06 ohm are connected in parallel and a resistance of 0.24 ohm is connected across it, determine (*i*) current delivered to the load resistor (*ii*) current delivered by each cell (*iii*) voltage across the load resistor and (*iv*) terminal voltage across each cell. (*Ans.* 6A, 1A, 1.44 V, 1.44 V)
3. Two lead-acid batteries each of 12V are connected in parallel to supply a load resistor of 1.2 ohm. Each battery consists of six cells each having 2.0 V, 0.4 ohm internal resistance, determine the current supplied by each battery. (*Ans.* 2.5A)

Type-III: Test Questions

1. How will you differentiate primary and secondary cells. Define emf, terminal voltage and internal resistance of an electric cell.
2. What is the necessity of grouping of electric cells? Derive a relation for the current delivered by a series-parallel grouping of cells. Show that the current delivered by a series-parallel grouping will be maximum when load resistance is equal to the internal resistance of the grouping.
3. Explain with neat diagram, the constructional details of a lead-acid battery.
4. Explain the working of a lead-acid battery with the help of chemical reactions.

3.11. CAPACITY OF A BATTERY

The quantity of electricity which a battery can delivery during single discharge until its terminal voltage falls to 1·8V/cell is called the **capacity** *of a battery.*

The capacity of a battery or cell is commercially expressed in ampere-hour and is generally denoted by A-H.

Capacity of a battery or cell $= I_d T_d$ ampere-hour

where, I_d = discharging current in ampere

T_d = discharging time of battery or cell in hour.

3.12. CHARGE INDICATIONS OF A LEAD-ACID BATTERY OR CELL

The charge condition of a battery is usually checked by checking the specific gravity of the electrolyte (H_2SO_4). For a fully charged battery, the specific gravity of H_2SO_4 is 1.28 to 1.29. However, when the specific gravity falls below 1.15, the battery is fully discharged. In fact, to increase the life of battery, it should be recharged when the specific gravity of the electrolyte is found to be less than 1.18. The values of specific gravity for different condition of charge are given below :

Specific gravity	*Condition*
1.280 to 1.290	100% charged
1.230 to 1.250	75% charged
1.190 to 1.200	50% charged
1.150 to 1.160	25% charged
below 1.130	fully discharged

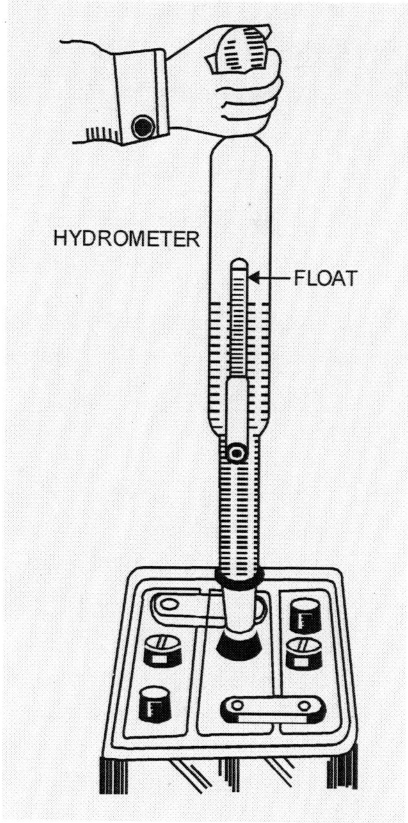

Fig. 3.10

To check the specific gravity of H_2SO_4, an instrument called *hydrometer* is used (see fig. 3.10) which works on Archmedeies principle. In common use, the decimal point is omitted from the value of specific gravity *i.e.* 1.280 specific gravity is spoken as 1280 and so on.

However, the state of a battery can also be checked by checking.

(*i*) *Voltage:* When the terminal voltage of the battery on load is 2.1 to 2.5 V per cell, the battery is said to be fully charged. Whereas, when the voltage of battery falls to below 1.8 V per cell, the battery is considered to be fully discharged and it is immediately put on charging.

(*ii*) *Colour of plates*: When a lead-acid cell or battery is fully charged, its anode is PbO_2 which is *chocolate brown* in colour and cathode is of *Pb* which is *grey* in colour. However, when the battery is fully discharged, both the plates attain $PbSO_4$ as active material which is whitish in colour.

3.13 CHARGING OF LEAD-ACID BATTERY

Whenever, terminal voltage of a battery falls below 1.8V per cell, it is put under recharging. The following points must be kept in mind while charging a battery:

(*i*) Only a d.c. voltage source is applied for recharging.

(*ii*) Ensure that + ve terminal of the source is connected with + ve terminal of the battery and negative with negative.

(*iii*) The charging voltage of the source should be approximately 2.5 V per cell.

(*iv*) The charging current should be maintained approximately 1A per positive plate of the battery per cell. For instant, if a cell of the battery contains 17 plates, it will have 8 positive plates. Thus, the charging current for such a battery will be 8A.

Note: An over current may produce excessive heat and damage the battery. Generally, the following two methods are employed for charging a battery.

1. Constant current method

In this method, the charging current supplied to the battery is kept constant throughout the charging period by adjusting the value of variable resistor R form R_1 to R_2 as shown in fig. 3.11 (a).

At the beginning of the charging, current supplied to the battery,

$$I = \frac{V - nE_1}{R_1 + nr}$$

where, n = No. of cells in series; r = internal resistance of each cell;
E_1 = e.m.f. of each cell in the beginning; R_1 = initial value of series resistor.

$$R_1 = \frac{V - nE_1}{I} - nr$$

As charging starts, the e.m.f. developed in each cell starts increasing which decreases the circuit current. To maintain the current constant, the value of series resistor is decreased gradually. Finally, when the battery is fully charged, let the e.m.f. developed per cell be E_2 and the value of series resistance be R_2. Then

$$R_2 = \frac{V - nE_2}{I} - nr$$

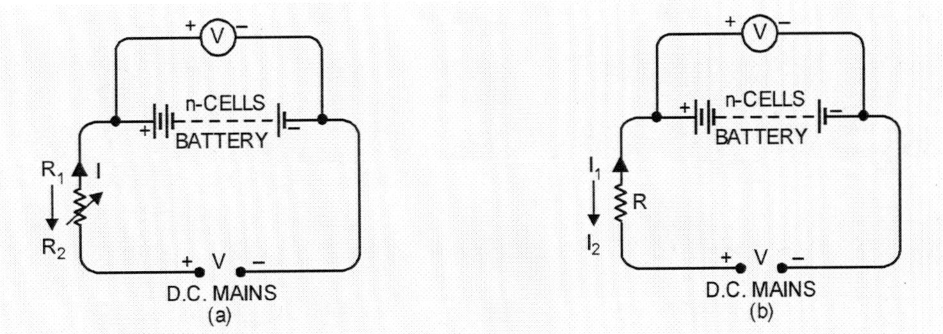

(a) (b)

Fig. 3.11

2. Constant voltage method

In this method, the supply voltage for charging the battery is kept constant. A fixed resistance R is connected in series with the battery, as shown in fig. 4.11 (b) to limit the current supplied to the battery in the beginning. In this case, the battery draws heavy current I_1 in the beginning which reduces to I_2 finally when the battery is fully charged.

$$I_1 = \frac{V - nE_1}{R + nr}; \; I_2 = \frac{V - nE_2}{R + nr}$$

This method of battery charging is applied commercially as total time required for complete charging is far less (nearly half) than that for constant current method. Moreover, no attendant is required to vary the series resistance.

Example 3.4. *A battery of 50 cells in series is charged through a resistance of 4 W from a 230V supply. If the terminal voltage per cell is 2V and 2.7V respectively at the beginning and at the end of the charge, calculate the charging current at the beginning and at the end of the charge.*

Solution: Charging current at the beginning,

$$I_1 = \frac{V - nE_1}{R + nr} = \frac{V - nE_1}{R} \quad \text{(since } r = 0\text{, not given)}$$

where, $V = 230\text{V}$, $n = 50$; $E_1 = 2$ V; $R = 4\ \Omega$;

$$\therefore \quad I_1 = \frac{230 - 50\times 2}{4} = \frac{130}{4} = \mathbf{32{\cdot}5\ A}\ (Ans.)$$

Charging current at the end,

$$I_2 = \frac{V - nE_2}{R} = \frac{230 - 50\times 2.7}{4} \quad (\because\ E_2 = 2.7\text{ V})$$

or $$I_2 = 95/4 = \mathbf{23{\cdot}75\ A}\ (Ans.)$$

Example 3.5. *A secondary battery consists of 6 cells in series, each having an e.m.f. of 1·8 V at discharge and 2·5 V when fully charged. The internal resistance of each cell is 0·1 W. If a charging supply of 24V is available, calculate the value of variable series resistance required to maintain the charging current at a constant value of 10A throughout the charging period.*

Solution: Value of series resistance in the beginning.

$$R_1 = \frac{V - nE_1}{I} - nr$$

where $V = 24$ V; $n = 6$; $E_1 = 1{\cdot}8$ V; $I = 10$ A; $r = 0{\cdot}1\Omega$

$$R_1 = \frac{24 - 6\times 1{\cdot}8}{10} - 6\times 0{\cdot}1 = \mathbf{0{\cdot}72\Omega}\ (Ans.)$$

Value of series resistance when the battery is fully charged.

$$R_2 = \frac{V - nE_2}{I} - nr; \text{ (where } E_2 = 2{\cdot}5\text{ V)}$$

$$= \frac{24 - 6\times 2{\cdot}5}{10} - 6\times 0{\cdot}1 = \mathbf{0{\cdot}3\ \Omega}\ (Ans.)$$

Learning outcome: *The value of series resistor has to be regulated from 0.72 Ω to 0.3 Ω to maintain the charging current at 10 A.*

3.14 CARE AND MAINTENANCE OF LEAD-ACID BATTERIES

The average life of a lead-acid battery is two to four years depending upon its manufacturing qualities and technique. However, to obtain longer life and efficient service, the following points must be kept in view:

1. The battery should not be allowed to use when the e.m.f. of the battery falls to 1·8 V per cell. Otherwise, the lead sulphate of the plates partly changes to non-active lead sulphate and reduces the life of the battery.
2. The specific gravity to the electrolyte should not be allowed to falls below 1.15.
3. The battery should never be left standing in a discharged condition, otherwise sulphation will occur and the battery cells are permanently damaged.
4. When not in use, the battery must be fully charged and stored in a cool and dry place.
5. Great care should be taken that the acid used as electrolyte should not contain any substantial impurity. It should be colourless when viewed through a 12 cm column.
6. The electrodes must remain completely immersed in the electrolyte, preferably the level of electrolyte should always be about 10 mm above the electrodes.
7. Whenever the level of the electrolyte decreases due to evaporation or gassing, distilled water should be added so as to keep the same concentration of electrolyte.
8. The battery should be charged and discharged at low rate so that its temperature may not rise above 45 °C. The high temperature may buckle the plates and damage the separators and the battery may be totally damaged.
9. The battery terminals should never be short circuited.
10. While charging the polarity must be checked carefully.
11. The room where the batteries are charged should be well ventilated as the atmosphere near the batteries would be charged with corresponding acid fumes.
12. The flames must be kept away form the vent of the battery, otherwise hydrogen and oxyzen produced with in the battery cells may get fire.
13. The battery terminals should always be kept clean and periodically greased with vaseline to prevent corrosion.

3.15 APPLICATIONS OF LEAD-ACID BATTERIES

Lead-acid batteries have innumerable commercial applications. Some of the important application are given below:

1. Used in automobiles for starting and lighting.
2. For lighting on steam and diesel railway trains.
3. Used at generating stations and sub-stations for operation of protective devices and for emergency lighting.
4. Used at telephone exchanges.
5. Used for emergency lighting at important places such as hospitals, theaters, banks etc.
6. Used for lighting purposes in remote rural areas.

TEST YOUR LEARNING - 2

Type-I: Short Answer Questions (*Section–A: 2 marks each*)

1. What do you mean by capacity of a battery? What are its units?
2. What are the charge indications of a lead-acid battery?
3. What precautions are to be taken while charging a battery?
4. Mention five important applications of a lead-acid battery.

Type-II: Numerical for Practice

1. A battery of 60 cells in series is charged through a resistance of 4 Ω from a 230V supply. If the terminal voltage per cell is 2V and 2.7V respectively at the beginning and at the end of the charge, calculate the charging current at the beginning and at the end of the charge.

   (*Ans.* 27.5A, 17A)

2. A secondary battery consists of 6 cells in series, each having an e.m.f. of 1·8 V at discharge and 2.4 V when fully charged. The internal resistance of each cell is 0·1 W. If a charging supply of 24V is available, calculate the value of variable series resistance required to maintain the charging current at a constant value of 10A throughout the charging period.

   (*Ans*. 0.72 W, 0.36 W)

Type-III: Test Questions

1. How hydrometer is used to determine the charged condition of a lead acid cell? Explain.
2. Explain the commercial method of charging a lead-acid battery.
3. Write a note on care and maintenance of a lead-acid battery.

Fig. 3.12: *Li-iron battery* (used in various mobile phones)

3.16 LITHIUM-IRON BATTERY

A lithium-ion battery or Li-ion battery (abbreviated as LIB) is a type of secondary battery and can be recharged. During discharging lithium ions move from the negative electrode to the positive electrode and back when charging.

In Li-ion batteries intercalated lithium compound is used as one electrode material, whereas, metallic lithium is used as electrode

material in a non-rechargeable lithium battery. The electrolyte, which allows for ionic movement, and the two electrodes are the constituent components of a lithium-ion battery or cell.

Chemistry, performance, cost and safety characteristics vary across LIB types. Handheld electronics mostly use LIBs based on lithium cobalt oxide ($LiCoO_2$), which offers high energy density but presents safety risks, especially when damaged. Lithium iron phosphate ($LiFePO_4$), lithium ion manganese oxide battery ($LiMn_2O_4$, Li_2MnO_3, or LMO), and lithium nickel manganese cobalt oxide ($LiNiMnCoO_2$ or NMC) offer lower energy density but longer lives and less likelihood of unfortunate events in real-world use (e.g., fire, explosion, etc.). Such batteries are widely used for electric tools, medical equipment, and other roles. NMC in particular is a leading contender for automotive applications. Lithium nickel cobalt aluminum oxide ($LiNiCoAlO_2$ or NCA) and lithium titanate ($Li_4Ti_5O_{12}$ or LTO) are specialty designs aimed at particular niche roles. The newer lithium–sulfur batteries promise the highest performance-to-weight ratio.

Lithium-ion batteries are common in domestic electronic devices. They are one of the most popular types of rechargeable batteries for portable electronic equipment. It is because of its high energy density, tiny memory effect and low self-discharge. LIBs are also growing in popularity for military, battery electric vehicle and aerospace applications.

If a lithium-ion cell or battery is charged too quickly, it may cause a short circuit, leading to explosions and fires. Because of these risks, testing standards are more stringent than those for acid-electrolyte batteries.

Shapes of lithium-ion battery pack.

Li-ion cells (as distinct from entire batteries) are available in various shapes, which can generally be divided into four groups:

- Small cylindrical (solid body without terminals, such as those used in older laptop batteries)
- Large cylindrical (solid body with large threaded terminals)
- Pouch (soft, flat body, such as those used in cell phones and newer laptops; these are also called lithium polymer batteries)
- Prismatic (semi-hard plastic case with large threaded terminals, such as vehicles' traction packs)

Electrochemistry

A lithium-ion battery or cell consists of negative and positive electrodes and the electrolyte providing a conductive medium for lithium ions to move between the electrodes.

Electrical energy flows out from or in to the battery when electrons flow through an external circuit during discharging or recharging respectively.

Both electrodes allow lithium ions to move in and out of their structures with a process called *insertion (intercalation)* or *extraction (deintercalation)*, respectively. During discharge, the (positive) lithium ions move from the negative electrode (anode) to the positive electrode (cathode) through the electrolyte while the electrons flow through the external circuit in the same direction. When the cell is charging, the reverse occurs with the lithium ions and electrons move back into the negative electrode in a net higher energy state.

The cell's energy depends upon the quantity of lithium. Each gram of lithium represents 13,901 coulombs. At 3 V, this gives 41.7 kJ per gram of lithium, or 11.6 kWh per kg. This is a bit more than the heat of combustion of gasoline, but does not consider the other materials that go into a lithium battery and that make lithium batteries many times heavier per unit of energy.

SUMMARY

1. *Electric cell*: A device in which chemical energy is converted into electrical energy.
   (*i*) *Primary cell*: The cells in which chemical action is not reversible.
   (*ii*) *Secondary cell*: The cells is which chemical action is reversible.
2. *Battery*: A number of cells connected in series, placed in single container is called a *battery*.
3. *Lead-acid cell (Working)* – Chemical action
   (*i*) During discharging

   At negative plates – $SO_4^{-} - 2e \longrightarrow SO_4$ (radical)

   $Pb + SO_4 \longrightarrow PbSO_4$

   At positive plates – $2H^{+} + 2e \longrightarrow 2H$

   $PbO_2 + 2H \longrightarrow PbO + H_2O$

   $PbO + H_2SO_4 \longrightarrow PbSO_4 + H_2O$

   (*ii*) During recharging:

   At positive plates – $SO_4^{-} - 2e \longrightarrow SO_4$ (radical)

   $SO_4 + H_2O \longrightarrow H_2SO_4 + O$

   $PbSO_4 + O + H_2O \longrightarrow PbO_2 + H_2SO_4$

   At negative plates – $2H^{+} + 2e \longrightarrow 2H$

   $PbSO_4 + 2H \longrightarrow Pb + H_2SO_4$
4. *Capacity of battery:* The quantity of electricity (in Ah) which a battery can deliver in single discharge is called its capacity.
5. *Charge indications of a lead-acid cell or battery:*
   (*i*) Specific gravity of electrolyte: 1.280 to 1.290 – 100% charged
   (*ii*) Voltage: more than 2V – fully charged; below 1·8 V – fully discharged.
   (*iii*) Colour of plates: anode – chocolate brown and cathode – grey fully charged; both whitish – fully discharged.
6. *Lithium-ion battery or Li-ion battery:* A lithium-ion battery or cell consists of negative and positive electrodes and the electrolyte providing a conductive medium for lithium ions to move between the electrodes.

SELF ASSESMENT
TEST YOUR PREPARATION

OBJECTIVE TYPE QUESTIONS

Type-I: Fill in the blanks:

1. The device which converts chemical energy into electrical energy is called ____________.

2. The specific gravity of the electrolyte is measured to check the charge condition of a ____________ battery.
3. When a lead-acid cell is fully charged, the specific gravity of the electrolyte will be ____________.
4. When a lead-acid cell is fully discharged, the colour of its positive plates will be ____________.
5. The capacity of a battery is measured in ____________.
6. For battery charging ____________ supply is required.
7. ____________ is used to measure the specific gravity of electrolyte of a lead-acid battery.
8. A lead-acid cell should not be discharged beyond ____________ volt.

Type-II: Multi-choice questions:

1. **The cell in which chemical action is not reversible is known as**
   (*a*) secondary cell (*b*) primary cell (*c*) voltaic cell (*d*) Edison cell
2. **The magnitude of e.m.f. developed in a cell depends upon**
   (*a*) nature of plates material and electrolyte;
   (*b*) size of plates and quantity of electrolyte;
   (*c*) both *a* and *b*;
   (*d*) none of above.
3. **A 2.2 V lead-acid cell**
   (*a*) is a primary cell (*b*) is a secondary cell
   (*c*) is a dry cell (*d*) has maximum current rating of about 150 mA
4. **The Edison cell**
   (*a*) is primary cell (*b*) uses H_2SO_4 as electrolyte
   (*c*) has 2.9 V output (*d*) contains nickel and iron electrodes.
5. **The capacity of a cell depends upon**
   (*a*) nature of plates material and electrolyte
   (*b*) size of plates and quantity of electrolyte
   (*c*) both *a* and *b* (*d*) none of above
6. **A 12 V lead-acid battery used in a car contains**
   (*a*) 10 cells connected is series (*b*) 10 cells connected in parallel
   (*c*) 6 cells connected is parallel (*d*) 6 cells connected in series.
7. **To check the specific gravity of an electrolyte, the name of the instrument used is**
   (*a*) hydrometer (*b*) lactometer (*c*) barometer (*d*) voltameter
8. **When lead-acid battery is fully charged.**
   (*a*) The colour of positive plates will be chocolate brown

(*b*) The colour of negative plates will be whitish

(*c*) the specific gravity of electrolyte will be 1.15.

(*d*) All above

NUMERICALS FOR PRACTICE

1. A lead-acid cell maintain a constant current of 2A for 15 hour before its terminal voltage falls to 1·8 V. What is the capacity of the cell? (**Ans.** *30 Ah*)
2. Fifty cells of 1·8 V each, having an internal resistance of 0·01 ohm are charged form 120 V d.c. supply. Determine the charging current and power drawn form the supply mains at the beginning if the external resistance the circuit is 1·25 ohms. What will be the current drawn from the supply mains when the cells are charged to 2.4 V each.

   (**Ans.** *17.143 A, 2057 W; zero*)
3. Four lamps in parallel, each requiring 0·5 A at 8V are to be lighted with a 12 V battery of internal resistance 0·5 Ω. Calculate the value of additional resistor required and draw the circuit diagram showing the method of connection. (**Ans.** *1·5 Ω*)

VIVA-VOCE/REASONING QUESTIONS

1. Usually, a large space is left between the plates and the bottom of the battery container, why ?
2. Large number of holes are provided in the separator (rubber sheet) placed between the positive and negative plates of lead acid battery, why?
3. While checking the conditions of battery it is seen that the float dips at different levels at different specific gravity of electrolyte, why?

SHORT ANSWER TYPE QUESTIONS

1. How will you define an electric cell?
2. What are the essentials for forming an electric cell?
3. How will you differentiate between primary and secondary cells?
4. What do you mean by emf and terminal voltage at a cell?
5. What is the necessity of grouping of electric cells?
6. Name the parts of a lead-acid battery. What is the function of container.
7. Why large number of holes are provided in the separator.
8. Write down the chemical reactions that took place in the lead acid battery during discharging.
9. Write down the chemical reactions that took place in a lead-acid battery during recharging.
10. What do you mean by capacity of a batter, what are the factors on which it depends?
11. Mention the charge indications of a lead-acid battery.

12. Why and how do we recharge the lead acid battery?
13. How do we care a lead-acid battery? Write five major points.
14. What are the applications of lead-acid batteries?

TEST QUESTIONS

1. What are the essentials of an electric cell? How the cells are classified?
2. Describe the construction and action of a lead-acid cell giving the nature of the chemical changes that occur at the electrodes both during charging and discharging.
3. Explain the construction and working of a lead-acid storage battery.
4. State the precautions to be observed in the use of a lead-acid cell.
5. Explain constant current and constant voltage method of charging a battery. Which method is employed commercially and why?
6. Give atleast five important application of a lead-acid battery.
7. Write a short note on Lit-iron cell or battery.

ANSWERS TO OBJECTIVE TYPE QUESTIONS

(i) Fill in the blanks:

1. an electrical cell **2.** lead-acid **3.** 1.28 to 1.29 **4.** Whitish
5. A-H **6.** d.c. **7.** Hydrometer **8.** 1.8

(ii) Multi-Choice Questions:

1. (*b*) **2.** (*a*) **3.** (*b*) **4.** (*d*) **5.** (*b*)
6. (*d*) **7.** (*a*) **8.** (*a*)

CHAPTER

4 LIGHTING EFFECT OF ELECTRIC CURRENT

TOPICS TO BE DISCUSSED

Lighting effect of electric current, filaments used in lamps, and Tube-light, LED, their working and applications.

LEARNING OBJECTIVES

After the completion of this **chapter,** *the students/readers will be able to understand:*

1. What is light energy and what are its sources?
2. Natural and artificial sources of light.
3. How electrical energy is used to produce light?
4. What are the various types of electric lamps? What are their characteristics and applications?
5. What should be the characteristics of material used to form filament of an incandescent lamp and how it is designed?
6. Why sodium vapour lamps and mercury vapour lamps are preferred for street lighting.
7. WHY *CFLs* are preffered over ordinary fluorescent tubes?
8. What are *LED* lamps, why these are preferred in back-lights of cars and signals placed at road-crossings.

INTRODUCTION

It is a known fact that almost all human activities ultimately rely upon light. Light is obtained from different sources.

Sources of Light

The light sources may be classified as natural sources and artificial sources:

(*a*) *Natural light sources*: The objects that are found in nature and are capable of emitting visible light are called *natural light sources, e.g.* sun, stars, meteors and meteorites. These objects either emit light from their own surface or reflect it.

(*b*) *Artificial light sources*: Any source of light that is manufactured by human being is called *artificial light source e.g.* candles, flame operated lamps, electric lamps such as incandescent tungsten filament lamps, fluorescent tubes, helogen lamps, low pressure mercury vapour lamps, compact fluorescent lamps, *LED* lamps etc.

Sun is the prime natural source of light but we can't do without artificial lighting. Artificial lighting can be produced by mechanical or electric lamps. In the beginning, mechanical lamps were used for lighting and many developments were made in these lamps to improve their efficiency. But slowly and steadily these lamps have become obsolete and are being replaced by the electric lamps.

At present almost all the lighting is produced electrically because of its following advantages

(*i*) cleanliness (*ii*) ease of control

(*iii*) economic consideration (*iv*) easy to handle

(*v*) connivent fittings (*vi*) Steady output

(*vii*) better reliability (*viii*) suitable for special applications etc.

Electrical lighting is not only employed in day-to-day life but it is also used for decorative purposes, advertising, street lighting, traffic control, medical field etc. In the industry, electric lighting plays an important role in increasing production, reducing workers fatigue, protecting workers health, eyes and nervous system and reducing accidents.

The usual method of producing artificial light electrically consists, in raising a solid body or vapours to incandescence by passing electric current or applying electric potential.

In fact, artificial light has become part and particle of our life. Our activities would stop without it. To understand various factors related to electric lighting we shall go through the different topics related to lighting effecrs of electric current, in this chapter.

4.1 NATURE OF LIGHT

As we know, one methods to produce artificial light is by increasing the temperature of a solid body to incandescence. At the start when the temperature of the body (say a piece of iron) is raised above that of its surroundings, it starts emitting energy in the form of radiations in all directions, this form of energy is called heat. As the temperature is raised gradually, at a particular temperature, the body becomes visible having a dull red colour. At this stage the body emits light as well as heat radiations. The wave length of the radiating energy becomes smaller and smaller when the temperature is further raised, but their velocity remains constant at a value of 3×10^8 ms^{-1}. The waves having wave lengths varying from 0.75×10^{-6} m to 0.4×10^{-6} m are known as light waves. Velocity of these waves is related to the wave length and frequency by the relation $v = n$l where n is the frequency and l is the wavelength of the waves. Basically these waves are electromagnetic waves. However, the nature and behaviour, of these waves depend upon their wavelengths and are named accordingly as given in the following table:

S. No.	Wave Length in m	Nature of Waves
1.	550 m to 11 m	Radio waves
2.	11m to 3×10^{-4} m	Commercial broadcasting waves
3.	3×10^{-4} m to 0.75×10^{-6} m	Infrared or heat rays
4.	0.75×10^{-6} m to 0.4×10^{-6} m	Light rays
5.	0.4×10^{-6} m to 3×10^{-8} m	Ultra violet rays
6.	less than 3×10^{-8}m	X-rays, γ-rays etc.

The wave at S. No. 4 in the above table represent light waves varying from extreme red to extreme violet.

4.2 ELECTRICAL METHODS OF PRODUCING LIGHT

The different methods of producing light electrically may be classified in to three categories:

(*i*) By *establishing an arc between two electrodes* such as in carbon arc lamps, flame arc lamps and magnetite arc lamps.

(*ii*) By *passing electric current through a filament* thus raising its temperature to incandescence such as in tungsten filament incandescent lamps.

(*iii*) By *electric discharge through vapours or gases* such as in vapour lamps, fluorescent tubes and neon signs etc.

4.3 ELECTRICAL SOURCES OF LIGHT

Using the above phenomenons, a variety of electric lamps have been designed to meet with different requirements broadly, electric lamps may be classified as

1. Arc lamps
2. Incandescent or filament lamps
3. Gaseous discharge lamps such as:

(*a*) Sodium vapour lamp
(*b*) Mercury vapour lamp
(*c*) Fluorescent tube
(*d*) Halogen lamps
(*e*) Compact fluorescent lamps *CFLs*
(*f*) *LED* lamps

4.4 INCANDESCENT OR FILAMENT LAMPS

In the earlier stages arc lamps were used for general lighting. But these lamps were complicated in construction, unsafe due to harmful radiations and very poor in efficiency, therefore, these were replaced by filament lamps for general lighting.

Working Principles

All of us know that when a room heater is switched *ON* the supply, it gives out faint red light along with heat. The working temperature of a room heater is about 750°C and at this temperature the radiations are mostly in the red and infra red regions. The phenomenon where heat is accompanied by light is known as incandescence and the construction of the filament lamp is based upon this phenomenon.

When an electric current is passed through a fine metallic wire it raises the temperature of the wire. At low temperatures only heat is produced. As the temperature is increased heat as well as light radiations are radiated and the amount of light radiations go on increasing with the increase in wire temperature. An incandescent lamp essentially consists of a fine wire of highly resistive material placed in an evacuated glass bulb. Such lamps were first produced by Edition in 1879, in which he used carbonised paper filament. But a carbon filament cannot be operated at high temperatures, therefore, the efficiency of those lamps was very poor about 4 lumens/watt. To obtain higher efficiency, the filament has to be operated at high temperatures about 2500°C, Therefore, the material employed for the construction of filament must have certain properties as mentioned below:

Properties of the metal required for the construction of filament of an incandescent lamp

(*i*) *high melting point*; so that it can be operated at high temperature.

(*ii*) *high specific resistance*; so that it can produce more heat and have smaller size.

(*iii*) *low temperature coefficient*; so that filament resistance may not change at operating temperature.

(*iv*) *low vapour pressure;* so that it may not vaporise and

(*v*) *highly ductile* and very strong mechanically to withstand vibrations.

After carbon, which couldn't be operated successfully beyond 1850°C, osium and tantalum having melting point 2700°C and 2900°C respectively were tried. Later on when it became possible to draw tungsten into fine wires, it superseded all other filament materials. Tungsten filament can be safely operated at about 2400°C to 2750°C in vacuum, with an efficiency of about 10 to 14 lumen/watt.

Construction

Evacuated glass bulb filament lamp

A tungsten filament lamp is shown in fig.4.1 where a filament is enclosed in an evacuated glass bulb to prevent oxidation of the filament. But evacuation helps the evaporation resulting in blackening of the bulb from inside.

This not only reduces the efficiency of the lamp but it also reduces the life of the filament. It was observed later on that this difficulty can be overcome by filling the bulb with some chemically insert gas such as argon or nitrogen which led to discovery of gas filled lamps.

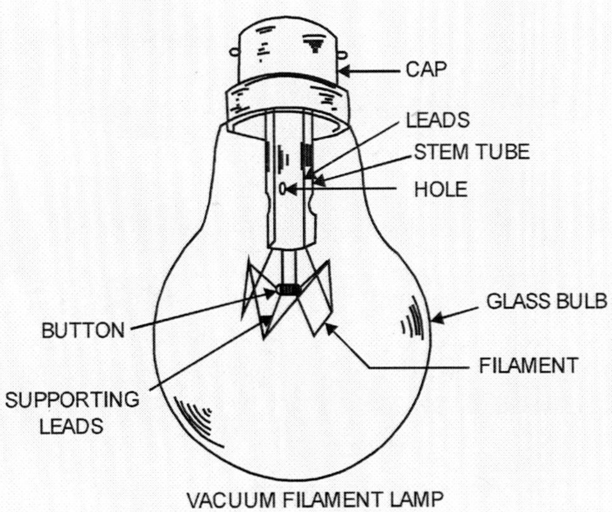

Fig. 4.1

Gas filled filament lamps

The basic working principle and constructing of this lamp is same as that of a vacuum filament lamp. In these lamps, to prevent evaporation of tungsten filament, the glass bulb is filled with some inner gas under pressure. The gases used for that purpose are generally argon and nitrogen. It is found that a mixture of 85% argon and about 15% nitrogen gave best results. The nitrogen is added to reduce the possibility of arcing within filament turns. By the use of inert gases, the tungsten filament can be safely operated at temperatures of about 2500°C to 2800°C, this also increases the efficiency of lamp. The introduction of gas, however, gave rise to another problem. It increases loss of heat from filament due to convection, thus again lowering the efficiency of lamp. Since the loss of heat directly depends upon the surface area of the filament, therefore, it can be reduced by decreasing surface area of the filament. This is achieved by either using single spiral (coiled) or coiled-coil filaments shown in fig.4.2.

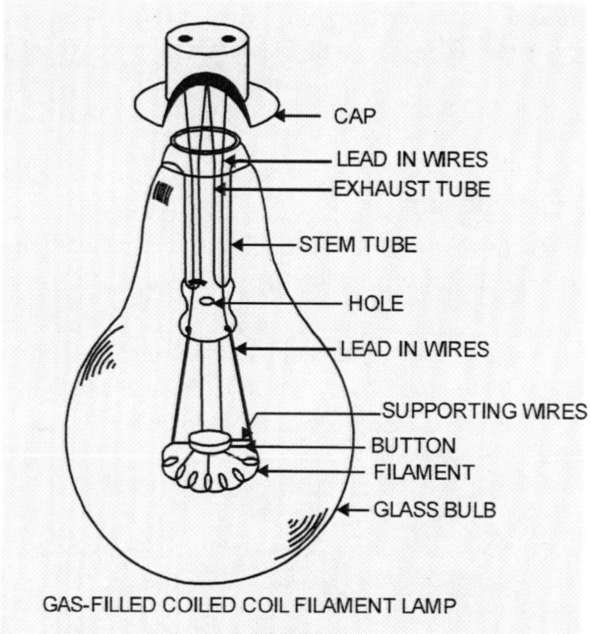

Fig. 4.2

In case of smaller wattage lamps, the loss of heat is comparatively more, so these are generally made vacuum type. Coiled coil filaments, shown in fig. 4.3 (*b* & *c*), are used for wattages up to 100 watt and for still higher wattages single coiled-coil filament are used. As the wattage of lamp increases the filament diameter also increases to carry larger currents. But as the diameter increases, the ratio of surface area to cross sectional area decreases. Hence lesser the cooling effect, higher the operating temperature and thus higher the efficiency.

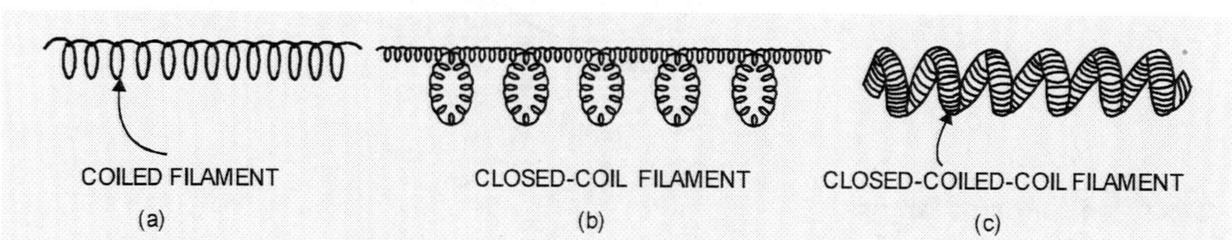

Fig. 4.3

The efficiency of a 40 watt lamp is about 10 to 12 lumen/watt, that of a 100 watt lamp is about 14 to 18 lumen/watt whereas the efficiency of 200 watt and more lamp may be 20 to 25 lumen/watt.

The normal life of incandescent lamps is 1000 working hours but it largely depends upon the operating voltage. If the operating voltage is below the rated value the life of the lamp increases on the other hand if the operating voltage is more than the rated value, the life of lamp decreases considerably.

Ageing effect

With the passage of time, its output and efficiency decreases due to evaporation of tungsten and blackening of inner surface of glass bulb.

Advantage

1. It operates at unity p.f.
2. Operates on standard supply voltage, no booster or no axiliary is required.
3. Available in various shapes and colours.
4. Good radiation characteristics in operating range.
5. Initial cost is low.

Disadvantages

1. It has poor efficiency due to which it consumes more power or energy for the same amount light output.
2. Operating life is less.
3. Heats-up the surroundings.

TEST YOUR LEARNING-1

Type-I: Short Answer Questions

1. What do mean by natural and artificial sources of light?
2. What is the range of wavelength of visible light?
3. Name major electrical sources of light.
4. What should be the properties of the metal required for the construction of filament of an incandescent lamp?

Type-II: Test Questions

1. Explain the construction and working of an incandescent lamp

4.5 FLUORESCENT TUBES

Construction

A fluorescent tube is a low pressure mercury vapour lamp. It consists of a glass tube 25 mm is diameter and 0.6 m, 1.2 m and 1.5 m in length. The tube contains argon gas at low pressure about 2.5 mm of mercury and a few drops of mercury. At the two ends, two electrodes coated with some electron emissive material are provided called filament of the tube.

In this low pressure mercury vapour lamp considerable amount of radiation is in the ultra-violet range. By coating the inside of the tube by phosphor these ultra-violet radiations are converted into visible light. Phosphors have the property of emitting visible light radiations when excited by ultra-violet radiations. Phosphors have definite characteristic colours. These are stable inorganic compounds and give a high output throughout the life of the lamp. The colours of fluorescence produced by various phosphors are given below:

	Phosphor	Colour
1.	Zinc silicate	Green
2.	Calcium tungstate	Blue
3.	Cadium Borate	Pink
4.	Calcium Holophosphate	Day light or white
5.	Magnesium tungstate	Bluish white

Working

A choke is connected in series with the tube which acts as a ballast and provides a high voltage impulse or surge for starting the glow in the tube. Whereas, during running (operating) condition the same choke absorbs some of the supply voltage and provides the remaining voltage of 110V across the lamp, which is sufficient for its operation. A capacitor is connected across the circuit to improve the power factor. A starting switch or starter is usually provided in the circuit which puts the electrodes directly across the supply mains at the time of starting so that electrodes may get heated and emit sufficient electrons.

Usually, Glow type starter is used to ignite the tube.

Working of Fluorescent tube using glow starter

The glow type starter is a voltage operated device and consists of two bimetallic electrodes enclosed in a glass bulb filled with a mixture of hydrogen and helium. Normally the contact between the bimetallic strips is open as shown in fig. 4.4.

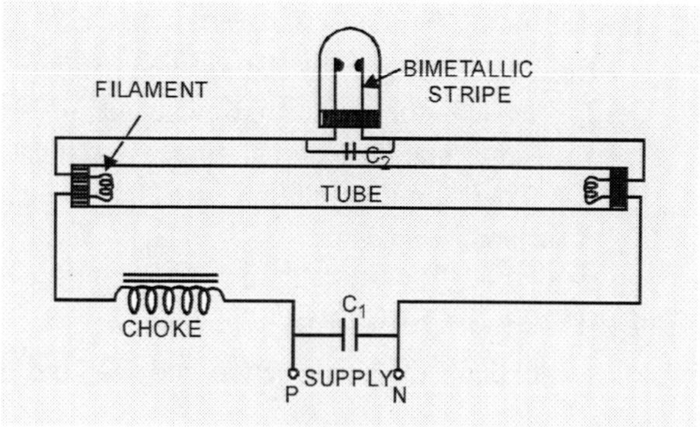

FIG. 4.4: Glow Type Starter Circuit

When supply is switched *ON*, the full mains voltage appears across the two bimetallic electrodes which causes a glow discharge in the glow switch. Due to this discharge a small amount of heat is produced which in turn causes the bimetallic strips to bend in such a way so as to make contact with each other. At this instant an appreciable amount of current flows through the choke, electrodes and strips, which raises the temperature of tube electrodes to incandescence. After one or two seconds the strips get cooled down and the contact is again opened.

The opening of the contact in series with the choke provides a momentary high voltage across the electrodes which is sufficient to start the discharge. The starter cannot glow after the tube has started operating because the potential available across the strips does not remain too high to cause glow discharge. During operating condition the starter does not consume any power, even if it is removed, it does not affected the working of the lamp.

4.6 COMPARISON BETWEEN TUNGSTEN FILAMENT LAMPS AND FLUORESCENT TUBES

The comparison between the two lamps is given below:

S.No.	Particulars	Tungsten Filament Lamp	Fluorescent Tubes
1.	*Effect of voltage fluctuation*	Voltage fluctuation has comparatively more effect on the light output.	Voltage fluctuation has comparatively small effect on light output as the variations in voltage are absorbed by the choke.
2.	*Effect of voltage on luminous efficiency*	Luminous efficiency per watt increases with the increase in applied voltage.	The luminous efficiency increases with the increase in applied voltage and length of the tube.
3.	*Effect of coloured light on luminous efficiency*	Luminous efficiency of coloured filament lamps is low because coloured glass absorbs light.	Their efficiency is high because coloured light is produced due to fluorescence.
4.	*Colour distortion*	It does not give light close to natural light, so colour rendering is defective.	It gives light close to nutural light and hence coloured objects can be properly seen.
5.	*Heat Radient*	Heat radiations are also present due to higher working temperature.	Heat radiations are neglegible due to low operating tempearture.
6.	*Brightness*	It brightness is more.	Its brightness is less.
7.	*Working life*	The average life of filament lamp is about 1000 working hours, but it varies with the working voltage. A slight increase in voltage may damage the lamp.	Its average life is about 6000 working hours. Life of fluorescent tube is not effected so much by variation in operating voltage but it depends upon the frequenc of starting.
8.	*Initial cost*	Initial cost per lamp is quite low.	Initial cost per tube is more.
9.	*Wiring cost*	For the same light output, a large number of lamps are reuired which results in a higher wiring cost.	For the same light output, smaller number of lutbes are required and hence wiring cost is low.
10.	*Effect of age on the output of the lamp*	The output of the lamp reduces with the passage of time.	The output reduces more rapidly as compared to filament lamp.
11.	*Manitenance cost*	Overall maintenance cost is low.	Overall maintenance cost is high.

4.7 COMPACT FLUROESCENT LAMPS (CFL)

The compact fluorescent lamps are becoming more and more popular now-a-days because of their low power consumption, low running cost, longer life, attractive look, soothing light and low maintenance.

These lamps are available in different sizes and designs as shown in fig. 4.5. These may have single rod, double rod, tripple rod or sprial rod. These lamps are available in different power rating *e.g.*, 5/7/9/11/18/24 watt, 220 V.

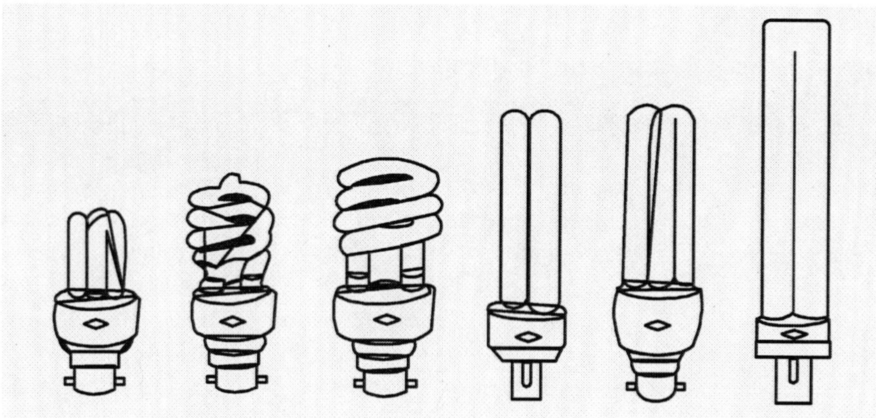

Fig. 4.5

Construction

Basically, it is a low pressure mercury vapour lamp having two electrodes coated with electron emissive meterial placed at the two ends of a glass tube. In the tube one drop of mercury and argon gass is filled at low pressure. The tube is coated internally with some fluroescent material in the form of powder. By mixing different powders the light of any desired colour (including day light) can be obtained. To improve its luminous efficiency a small amount of halogen is also added to the inert gas. An electronic balloast is used with the lamp to start the glow. The luminous efficiency of these lamps varies from 50 lm/W to 90 lm/W. The technical data of these lamps is given below:

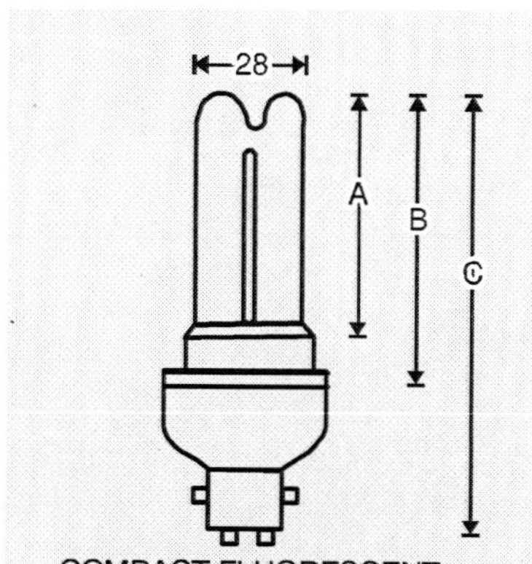

Fig. 4.6

Technical Data

Types of Lamps	Wattage (W)	Lumen (LM)	Lamp. Voltage (V)	Current (mA)	Dimenstion (mm)		
					A	B	C
CFL 5W	5	250	34	180	69	85	108
CFL 7W	7	400	47	175	91	115	138
CFL 9W	9	600	60	170	129	145	168
CFL 11W	11	900	91	160	199	215	238

Major Advantages of CFL

The following are the main advantages of *CFL* over the ordinary incandescent lamp.

(*i*) Low energy consumption

(*ii*) Low maintenance costs

(*iii*) It provides instant glow

(*iv*) It provides light without heating the surroundings

(*v*) Excellent colour rendering properties. Hence all colours look natural and desired ambience can be created in various interiors.

(*vi*) Low operational costs: The power consumption of *CFL* is 80% less than incandescent lamps.

(*vii*) The compact size opened up possibilities for exciting and aesthetically appealing new luminaries.

(*viii*) Average life of compact fluorescent lamp is 8000 working hours.

Applications

The compact size, lovely appearance, longer life, higher efficency, low running and maintenance cost, instant glowing power make these lamps suitable for all places where uniform illumniation (not flood light) is required. Some of such places are mentioned below:

Offices, shops, hotels, hospitals, conference halls, cinema halls, table lamp fittings, reception lobbies, residential buildings, emergency lighting systems etc.

4.8 LIGHT EMITTING DIODE (LED) LAMPS

Light Emitting Diode (*LED*) *is a semiconductor device. Basically, it is a chemical chip embedded in a plastic capsule. When it is energised by direct current it emits light.*

An *LED* lamp is a lamp in which large number of *LEDs* are assembled into a lamp (or light bulb) so that it can be used in lighting fixtures. These lamps are made to replace and screw-in incandescent or compact fluorescent light bulb holders. Such lamps are made with standard light bulb connections and shapes. Some of them arc shown in fig 4.7.

Fig. 4.7: LED lamps

LED's emit light in a particular direction but as these are very compact, they provide flexibiity in designing lighting fixtures and good control over the distribution of light with small reflectors or lenses. Moreover because of small size, control of the special distribution of illumination can be obtained as per the requirement. For special distribution of light, an *LED* array can be designed without any loss of efficiency.

Merits

Some of the important merits of these lamps are given below:

1. These lamps have longer lifespan about 40000 hour say about 8 year.
2. More power efficient as compared to all other electric lamps, almost 120 to 200 lumen per watt.
3. These do not have any warming-up time as *CFLs* need.
4. Light of various colours can be obtained without losing efficiency.
5. These are very light in weight.
6. They need very less maintenance.

Limitations

1. Initial cost is very high in comparison to other lamps.
2. Light emitted by *LEDs* is directional, they do not emit light in all directions.
3. *DC* is required for their operation, hence axiliaries are required.
4. These are adversely affected by high temperatures, therefore, for heat dissipation heat-sinks are used.

Applications

1. As indicating lamps in instruments, panal boards, switches, controls etc.
2. In trafic signals.
3. Automobile light fittings
4. Domestic, commercial and industrial places for illumination. These are available in the form of tubes having 2, 4 and 8 feet length.

In fact, these lamps can be used at all such places where other electric lamps are employed.

TEST YOUR LEARNING – 2

Type-I: Short Answer Questions

1. Why fluorescent tube is preferred over filament lamp?

2. How the colour of fluorescence is changed in fluorescent tube?

3. Why *CFL* is preferred over fluorescent tube?

4. Why *LED* lamp is preferred over *CFL*?

Type-II: Test Questions

1. Explain the construction of a fluorescent tube.
2. Make a coparison between filament lamp fluorescent tube.
3. Write a short note on (*i*) *CFL* (*ii*) *LED* lamp.

SUMMARY

1. *Sources of light:* The sources of light may be classified as
   (*i*) natural sources of light and (*ii*) artificial sources of light
2. *Natural sources of light:* The objects that are found in nature and are capable of emitting visible light are called natural light sources, *e.g.* sum, stars, meteors and meteorites.
3. Artificial sources of light: Any source of light that is manufactured by human being is called artificial light source *e.g.* candles, flame operated lamps, electric lamps such as incandescent tungsten filament lamps, fluorescent tubes, *LED* lamps etc.
4. *Advantages of electric lamps:*
   (*i*) cleanliness (*ii*) ease of control (*iii*) economic consideration
   (*iv*) easy to handle (*v*) connivent fittings (*vi*) steady output
   (*vii*) better reliability (*viii*) suitable for special applications etc.
5. *Nature of light:* Natural light is constituted by the electromagnetic waves. The wave length of light rays fall between 0.75×10^{-6} m to 0.4×10^{-6} m.
6. *Incandescent or filament lamp:* In these lamps, a filament is heated to the extent that it becomes incandescent (*i.e.*, it starts emitting light).
7. *Properties of the metal required for the construction of filament:*
   (*i*) high melting point (*ii*) high specific resistance
   (*iii*) low temperature coefficient (*iv*) low vapour pressure (*v*) highly ductle.
8. Characteristics of fluorescent tubes: A fluorescent tube is bassically a low pressure mercury vapour lamp having an efficiency of about 40 lumen/watt. It provides almost day light.
9. *Compact fluorecent lamps(CLFs):* These are available in different sizes and ratings. These are very efficient (about 80 lumen/watt).
10. *Advantages of CLFs:* (*i*) low operating cost (*ii*) low maintenance cost (*iii*) very long life (about 8000 working hours) (*iv*) high efficiency (about 80 lumen/watt) (*v*) instant glow (*vi*) compact is size (*vii*) attractive look.
11. *Light Emitting Diode (LED) Lamp:* It is a semiconductor device, a chemical chip is embedded in a plastic capsule which is energised by direct current and emits light.
12. *Merits and limitations of LED lamps:*

    Merits: (*i*) longer lifespan about 40000 hour say about 8 year, (*ii*) more power efficient almost 120 to 200 lumen per watt, (*iii*) no warming-up time, (*iv*) light of various colours can be obtained without losing efficiency, (*v*) very light in weight and (*vi*) need very less maintenance.

    Limitations: (*i*) Initial cost is very high. (*ii*) Light emitted by LEDs is directional, (*iii*) *DC* is required for their operation, hence axiliaries are required. (*iv*) These are adversely affected by high temperatures, therefore, for heat dissipation heat-sinks are used.

SELF ASSESMENT
TEST YOUR PREPARATION

OBJECTIVE TYPE QUESTIONS

Type-I: Fill in the blanks:

1. Artificial light can be produced by raising the temperature of a body to.........
2. Running cost of a fluorescent lamp is........... than incandescent lamp.
3. The efficiency of filament lamp is about........to........lumen/ watt.
4. The material of the filament in incandescent lamps is........
5. The normal life of tungsten filament lamp is........ working hours.
6. The fluorescent material used for coating the inside of fluorescent lamp is.............
7. The normal working life of fluorescent tube is........hours.
8. The function of choke in a fluorescent lamp circuit is.................
9. The efficiency of fluorescent tube is........lumen/watt.

Type-II: Multi-choice questions: *Select the correct answer in the following*:

1. The material of the filament in incandescent lamps is
   (*a*) copper (*b*) aluminium
   (*c*) tungsten (*d*) nicrome
2. The process taking place in a flurescent tube is called
   (*a*) gaseous discharge (*b*) secondary emission
   (*c*) phosphorescence (*d*) thermionic emission

SHORT ANSWER TYPE QUESTIONS

1. Explain the nature of light and show how an eye perceives differnet colours.
2. What do you mean by primary and secondary sources of light?
3. Why is tungsten selected as the filament material and on what factors does its life depend?
4. Explain the construction of an incandescent lamp with the help of a neat diagram.
5. Mention the properties of the material used for the construction of filament of an incandescent lamp.
6. Why coiled coil filament is preferred over coiled filament?
7. Draw the circuit of a fluorescent lamp with glow starter and explain its working.
8. Explain the function of a flow starter and choke used in the circuit of a fluorescent lamp.
9. Make a comparison between filament lamp and fluorescent tube.
10. State the advantages of fluorescent lamp over filament lamp.
11. Why *CFL*s have become so popular.
12. What are *LED* lamps? What are their characteristics and applications?

TEST QUESTIONS

1. Explain the nature of light.
2. Draw circuit diagram of a fluorescent lamp and explain the function of each component.
3. (i) Why is tungsten selected as the filament material and on what factors does its life depend?
   (ii) Why are condenser and choke used in fluorescent lamp?
   (iii) How are hard and long shadows avoided?
4. Explain, with diagrams, how visible light is produced from.
   (i) Incandescent lamp;
   (ii) Fluorescent tube.
5. (*a*) What is the use of coiled coil filament in an incandescen lamp? What are its advantages over ordinary coil filament?
   (*b*) State advantages and disadvantages of fluorescent lamps over filament lamps.
6. Explain the working of an *LED* lamp. What its advantages over *CFLs*?

ANSWERS TO OBJECTIVE TYPE QUESTIONS

(*i*) Fill in the blanks:

1. incandescence **2.** less **3.** 10, 14 **4.** tungsten
5. 1000 **6.** calcium holophosphate **7.** 7000
8. to give a blast at start **9.** 40

(*ii*) Multi-Choice Questions:

1. (*c*) **2.** (*a*)

CHAPTER

5 CAPACITORS

TOPICS TO BE DISCUSSED

Capacitor and its capacity, Concept of charging and Discharging of capacitors, Types of Capacitors and their use in circuits, Series and parallel connection of capacitors, Energy stored in a capacitor.

LEARNING OBJECTIVES

After the completion of this ***chapter,*** *the students/readers will be able to understand:*

1. What is a capacitor?
2. What is the meaning of capacitance of a capacitor?
3. What is the process of charging and discharging of a capacitor?
4. What are various types of capacitors?
5. How these capacitors are employed in various circuits?
6. Series, parallel and series-parallel connections of capacitors.
7. Impotance of these combinations.
8. Energy stored in a capacitor.

INTRODUCTION

Electrostatics is the branch of engineering or science that deals with the static (*i.e.* stationary) charges.

When a glass rod is rubbed with a silk cloth, the former attains positive charge and the later attains negative charge of equal magnitude. Since, glass rod and silk cloth both are insulators, they retain these charges. These charges are known as static charges and the process comes under the general heading of *electrostatics*.

These static charges can be stored in a device called *capacitor.*

In this chapter, we shall deal with various types of capacitors and their combination i.e. their series, parallel or series-parallel connections.

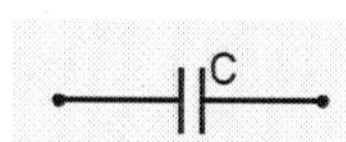

Fig. 5.1

5.1 CAPACITOR

*Two surfaces separated by an insulating material or *dielectric is called* a **capacitor** *or* **condenser**.

* Since current cannot pass through an insulator but an electric field can, therefore, an insulator is also called as dielectric.

Since this arrangement has the capacity to store electricity, it is named as *capacitor*. It is also known as condenser due to the fact that when p.d. is applied

across the conducting plates, the electric lines of force are condensed in the small space between them. A capacitor is generally named after the dielectric used *i.e.* air capacitor, paper capacitor, mica capacitor, ceramic capacitor etc. Depending upon the shapes of the conducting surfaces or plates, the capacitor may be in the form of parallel plates, concentric cylinders or spheres etc. The symbol of a capacitor is shown in fig. 3.25.

5.2 TYPES OF CAPACITORS

The capacitors are generally classified according to the dielectric used. The most commonly used capacitors are paper, mica, ceramic, electrolytic and air capacitors. These may be either fixed or variable type.

1. Paper Capacitors: Paper capacitors are the most common of all capacitors. For the construction of these capacitors two metal (aluminium or tin) foils separated by paper impregnated with a dielectric material such as wax, plastic or oil are rolled into a compact cylinder. Connecting leads are attached to the two metal plates. The entire cylinder is generally placed in a cardboard container coated with wax or encased in plastic. A tubular type paper capacitor is shown in Fig. 5.2.

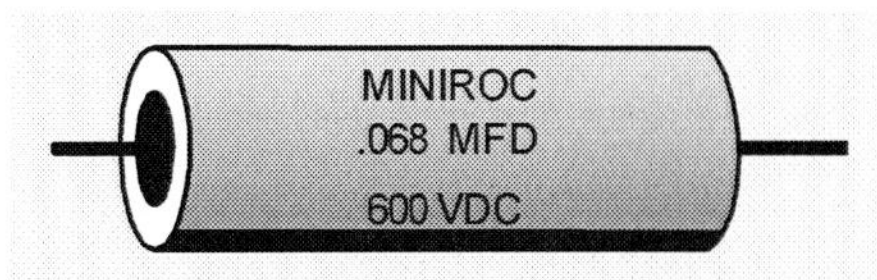

Fig. 5.2: Tubular Capacitor

Paper capacitors are available in a wide range of capacitance values and voltage ratings. Typical capacitance values ranging from 0.0001 μF to 1.0 μF having tolerance of $\pm$ 10 percent usually with voltage ranging from 200 to 10 000 volts and more are available.

2. Mica Capacitor: These capacitors consist of alternate thin sheets of metal (aluminium or tin) foils separated by thin mica sheets as shown in Fig. 5.3. Alternate metal sheets are connected together

and brought out as one terminal for one set of plates, while the opposite terminal connects to the other set of plates. The entire unit is generally encased in a plastic housing or moulded in bakelite case.

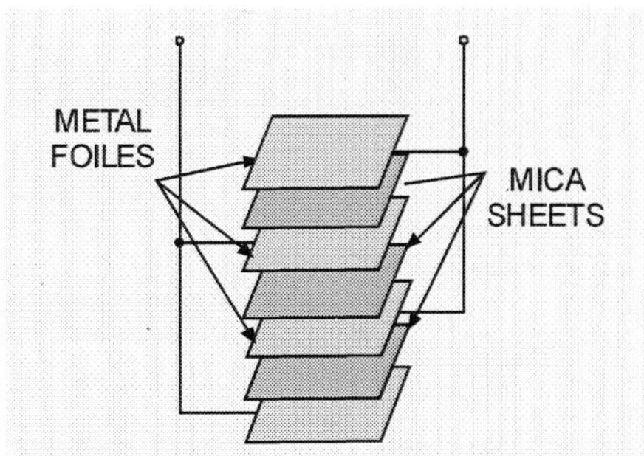

Fig. 5.3: Internal Construction of Mica Capacitors.

Mica capacitors are often used for small capacitance values ranging 50 to 500 πF with voltage ratings ranging from 200 to 1000 V.

3. Ceramic Capacitors: The ceramic is a dielectric material made from earth fired under extreme heat. Titanium oxide or several other types of silicates are used to obtain very high value of dielectric constant of ceramic material. The ceramic capacitors may be of disc type as shown in fig. 5.4 or tabular type. These capacitors are also available in other shapes.

Disc ceramic capacitors are commonly available in capacitance values ranging from 47 πF to 0.05 μF, with voltage ratings ranging from 200 to 1000 V.

In tubular ceramic capacitors, the inner and outer surfaces of a hollow ceramic tube are coated with silver and form the two plates of the capacitor. These capacitors are available in capacitance values ranging from 1 to 500 of. Special low-voltage, high capacitance ceramic capacitors are also available for use in transistor circuits.

4. Electrolytic Capacitors: An electrolytic capacitor contains two aluminium electrodes as shown in Fig. 5.5 (*a*). Between the two electrodes, absorbent gauze soaks up electrolyte (borax, phosphate or carbonate) to provide the required electrolysis that produces an oxide film (a molecular-thin layer of aluminium oxide) at the positive electrode when d.c. voltage is applied. The oxide film acts as an insulator and forms a capacitance between the positive aluminium electrode and the electrolyte in the gauze separator. The negative aluminium electrode simply provides a connection to the electrolyte. Usually, the metal itself act as the negative terminal of the capacitor. A tubular electrolytic capacitors are shown in Fig. 5.5 (*b*).

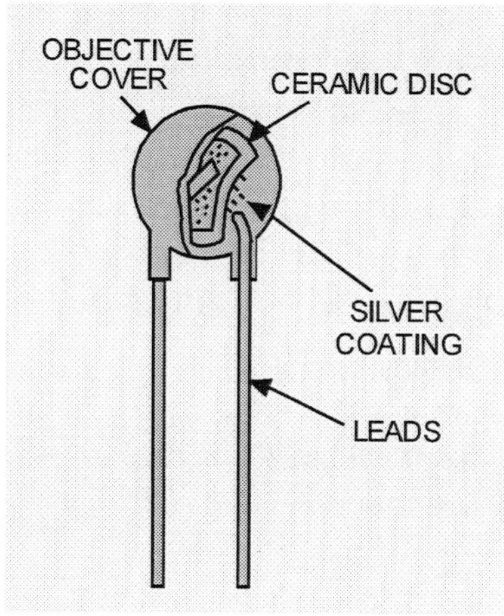

Fig. 5.4: Disc-Type Ceramic Capacitor

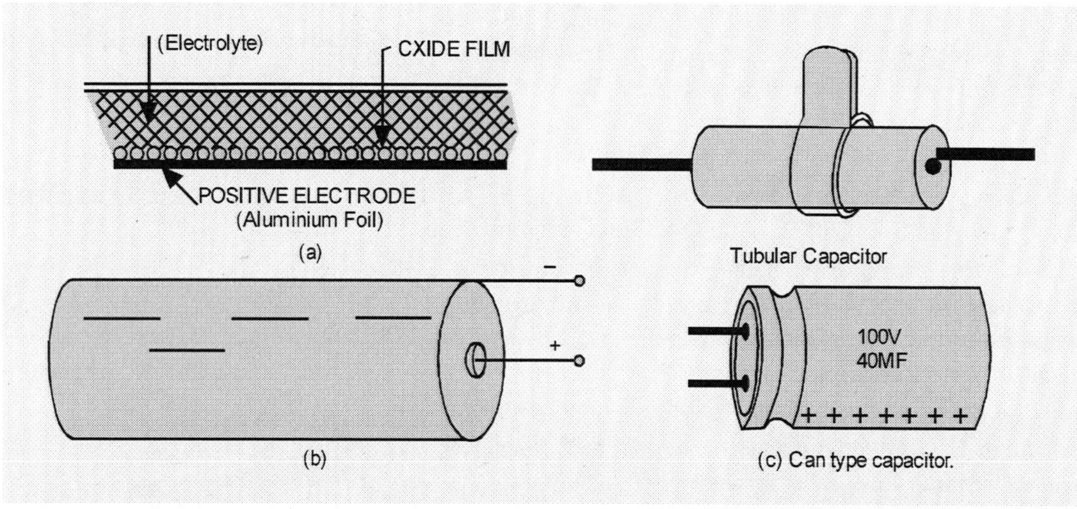

Fig. 5.5

Electrolytic capacitors have a high capacitance-to-size ratio since the aluminium oxide layer is molecular thin. These capacitors are available in capacitance values ranging from less than 1 μF to 10 000 μF or more. Common voltage ratings range from 1 to 700 volt.

These capacitors must be connected in the circuit as per polarity marked on the capacitor. If they are connected in opposite polarity, the reversed electrolysis forms gas in the capacitor. It becomes hot and may explode. Non-polarized electrolytic capacitors are also available which are applied in the starting winding to a.c. single-phase motors. Basically, a non-polarized electrolytic capacitor contains two capacitors, connected internally in *series-opposing* polarity. The capacitance is one-half either capacitance, but the oxide film is maintained.

Variable capacitors

The electronic circuits, in which frequency is to be changed as per requirement such as tuning circuits, variable capacitors are used. *Air-gang capacitor* is the most common variable capacitor. These capacitors are available in single, dual and three-section units. A two-section unit of air-gang capacitor is shown in Fig. 5.6.

Another type of variable capacitor is called *trimmer* (sometimes also called *paddler*). These capacitors are used in the electronic circuits where the variation of capacitance is not frequent. Once a setting is obtained to match the circuit, the capacitance is not to be changed after that. This capacitor consists of two or more metal (aluminium) sheets separated by sheets of mica or ceramic as shown in Fig. 5.6. The metal sheets are under spring tension so that as they are squeezed together by the tuning of a adjusting screw, their physical spacing and hence, trimmer capacitance is varied.

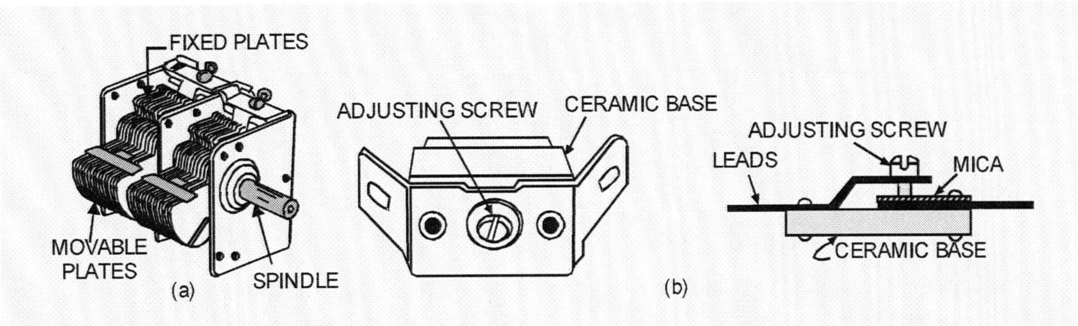

Fig. 5.6

5.3 CAPACITOR ACTION

When a d.c. supply is connected across a capacitor, it stores charge, The manner in which it stores charge is illustrated in Fig. 5.7. A parallel plate capacitor having plate A and B is connected across a battery of voltage V volt through a key K.

(*i*) When key K is open as shown in fig. 5.7 (*a*), the capacitor plates are neutral *i.e.* the plates are having no charge.

(*ii*) When key K is closed as shown in fig. 5.7 (*b*), the electrons from plate A are attracted by the positive terminal of the battery and reach at plate B through the battery. Thus the electrons detached from plate A start * piling up on plate B. As a result, plate A attains more and more positive charge and plate B attains more and more negative charge. This process is called charging of capacitor.

(*iii*) This flow of electrons continue from plate A to B through battery till the capacitor is charged to V volt (*i.e.* equal to the supply voltage). Once the capacitor is charged to V volts the flow of electrons ceases as shown in fig. 5.7 (*c*).

(*iv*) Now, an electrostatic field is established between the plates in the dielectric.

* The electrons cannot jump from plate B to A as there is insulating material between the plates.

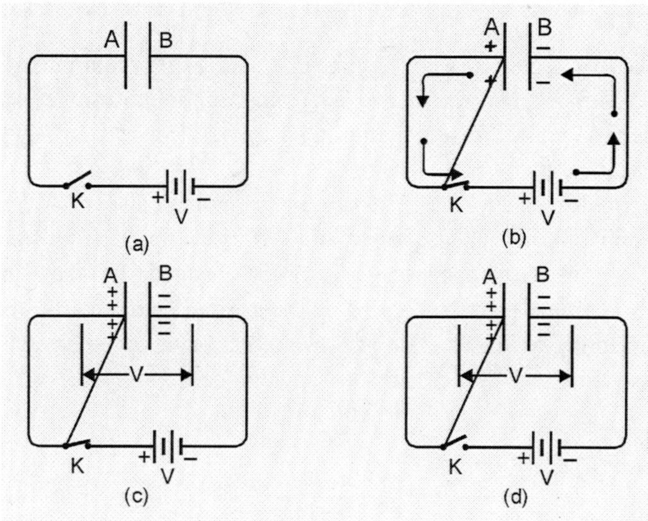

Fig. 5.7

5.4 CAPACITANCE

The capability of a capacitor to store charge is called its **capacitance.**

It has been explained above that charge Q stored by a capacitor is directly proportional to the p.d. V applied across it *i.e.*

$$Q \propto V \quad \text{or} \quad \frac{Q}{V} = \text{constant} = C$$

where, C is a constant called capacitance of a capacitor.

Hence, the charge on the capacitor plates per unit potential difference across the plate is called the **capacitance** *of the capacitor.*

The unit of capacitor is *__farad (F)__.

If $Q = 1$ coulomb; $V = 1$ volt

Then $C = 1\text{F}$

Hence, a capacitor is said to have a capacitance of **1 farad** *if a charge of 1 coulomb accumulates on each plate when a p. d. of 1 volt is applied across the plates.*

Farad is an extremely large unit of capacitance. Practically, the capacitors are manufactured having the capacitance of the order of microfarad (μF) and micro – microfarad (μμF) or picofarad (pF)

$$1\mu\,\text{F} = 10^{-6}\text{F} \quad \text{and} \quad 1\,\mu\mu\,\text{F} = 1\,\text{p}\,\text{F} = 10^{-12}\,\text{F}$$

5.5 DIELECTRIC CONSTANT OR RELATIVE PERMITTIVITY

A capacitor consists of two conducting surfaces (plates) separated by an insulating material called dielectric. When the capacitor is charged, the electrostatic field is established between the conducting plates. The concentration of electric lines of force increases with the presence of dielectric. Actually,

* Capacitance, $C = Q/V$; The unit of charge and voltage is coulomb and volt respectively. Therefore the unit of capacitance will be C/V. However, to honour Michel Faraday, the unit given to capacitance is Farad (symbol F).

the degree of concentration of electric lines of force between the plates depends upon the nature of dielectric.

The ability of a dielectric material to concentrate electric lines of force *between plates of the capacitor is called* **dielectric constant** *or* **relative permittivity** *of the material.*

Air has been assigned a reference value of dielectric constant (or relative permittivity) as *one*. The dielectric constant of all other insulating material is more than one.

Let V = potential difference applied across the capacitor

Q = charge on the capacitor plates with air as the insulating medium.

Then $$C_{air} = \frac{Q}{V}$$

Now, if mica is used as insulating medium in the same capacitor, it can hold a charge of $6Q$ coulomb with the same voltage V.

Then $$C_{mica} = \frac{6Q}{V} = 6\frac{Q}{V} = 6C_{air}$$

i.e. $$\frac{C_{mica}}{C_{air}} = 6 \quad (\text{i.e. dielectric constant of mica})$$

Hence, **dielectric constant** *or* **relative permittivity** *of a material may be defined as a ratio of capacitance of capacitor with that material filling the space between the plates to the capacitance of the same capacitor having air as filling material.*

TEST YOUR LEARNING-1

Type-I: Short Answer Questions

1. How do you distinguish between capacitor and capacitance.
2. What will happen if the plates of a charged capacitor are connected through a copper wire?
3. Can we give any desired charge to a capacitor?
4. Why mercury cannot be used as a dielectric in a capacitor?
5. Define capacitor, capacitance and farad.

Type-II: Numerical for Practice

1. What must be the capacitance of a capacitor which can store a charge of 0.015 C when a potential difference of 250 V is applied across it? (*Ans.* 60 μF)

Type-III: Test Questions

1. What is a capacitor? Classify capacitor on the basis of electrolyte used. Explain construction of a paper capacitor.
2. Explain the construction of an electrolyte capacitor with the help of neat sketches. Why polarity is marked on the terminals of an electrolytic capacitor? If the terminals of this capacitor are connected in opposite polarity, what will happen?
3. Explain capacitor action. What are the factors on which the capacitance of a parallel plate capacitor depends?

5.6 GROUPING OF CAPACITORS

The capacitors of different capacitance, rated at different voltages are available in the market. To obtain the required value of the capacitance, the capacitors may be connected in *parallel* (for larger value) in *series* (for smaller value) or in *series-parallel* grouping.

5.7 CAPACITORS IN SERIES

Consider three capacitors having capacitances C_1, C_2 and C_3 farad respectively connected in series as shown in fig. 5.8. When a p.d. of V volts is applied across the grouping, momentarily a charging current flows through the circuit which develops the same charge Q on each capacitor at different p.d. V_1, V_2 and V_3 respectively.

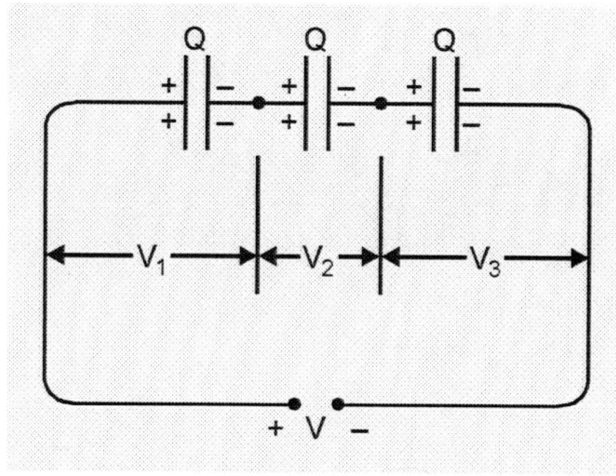

Fig. 5.8

Now, Supply voltage = Sum of p.d. across each capacitor

$$V = V_1 + V_2 + V_3 = \frac{Q}{C_1} + \frac{Q}{C_2} + \frac{Q}{C_3}$$

$$= Q\left(\frac{1}{C_1} + \frac{1}{C_2} + \frac{1}{C_3}\right)$$

$$\frac{V}{Q} = \frac{1}{C_1} + \frac{1}{C_2} + \frac{1}{C_3}$$

But Q/V is the total or effective capacitance C_T of the combination, as shown in fig. 5.9.

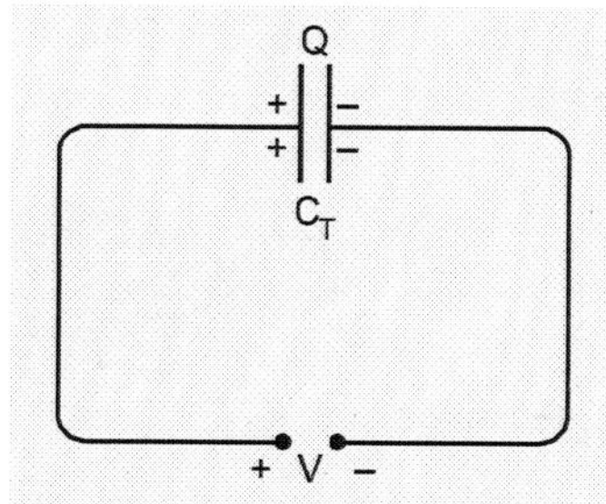

Fig. 5.9

$$\frac{1}{C_T} = \frac{1}{C_1} + \frac{1}{C_2} + \frac{1}{C_3}$$

Hence, when a number of capacitors are connected in series, the reciprocal of total capacitance is equal to the sum of reciprocal of their individual capacitances.

5.8 CAPACITORS IN PARALLEL

Consider three capacitors having capacitances C_1, C_2 and C_3 farad respectively connected in parallel as shown in fig. 5.10. When a p.d. of V volt is applied across the grouping, different charging current flows in each branch and the capacitors attain different charge *i.e.* Q_1, Q_2 and Q_3 respectively depending upon their capacitances. But the p.d. across each capacitor is the same.

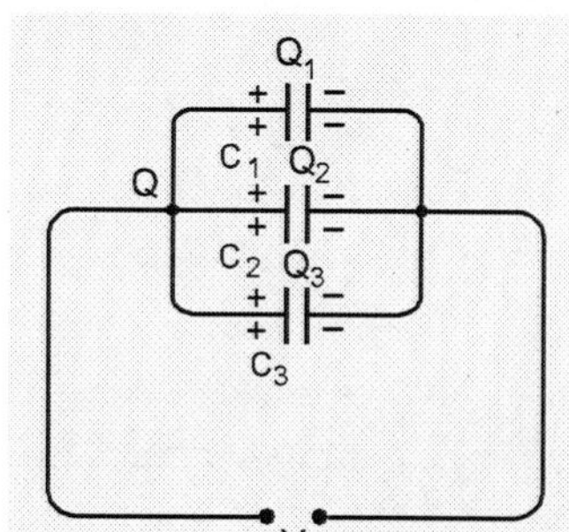

Fig. 5.10

Now, total charge shifted = Sum of the charge shifted on each capacitor

i.e.
$$Q = Q_1 + Q_2 + Q_3 = C_1V + C_2V + C_3V$$

$$= V(C_1 + C_2 + C_3)$$

or $$\frac{Q}{V} = C_1 + C_2 + C_3$$

But Q/V is the total or effective capacitance C_T of the combination as shown in fig. 5.11.

$$\therefore \quad C_T = C_1 + C_2 + C_3$$

Hence, when a number of capacitors are connected in parallel, the total capacitance is equal to the sum of their individual capacitances.

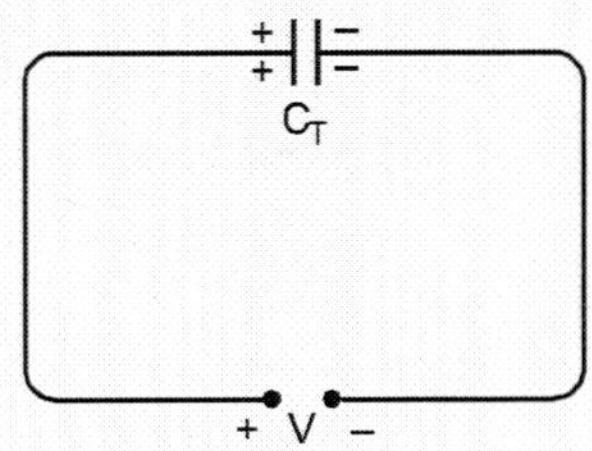

Fig. 5.11

5.9 CAPACITORS IN SERIES-PARALLEL

The series-parallel grouping of the capacitors is shown fig. 5.12 (*a*). In this type of circuits, first of all the effective value of each parallel grouping is determined and the circuit is reduced to series circuit as shown in fig. 5.12 (*b*). Then capacitance of the whole circuit is determined as given below:

$$C_{p1} = C_2 + C_3;\ C_{p2}$$

$$= C_4 + C_5 + C_6$$

$$\frac{1}{C_T} = \frac{1}{C_1} + \frac{1}{C_{p1}} + \frac{1}{C_{p2}}$$

Where C_T is the total or effective capacitance of the whole circuit as shown in fig. 5.12(*c*).

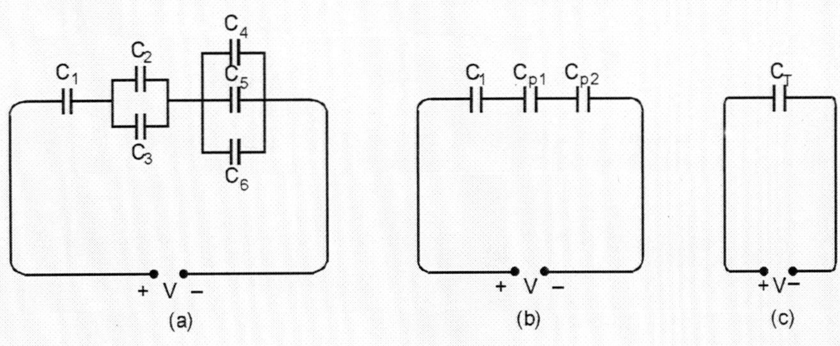

Fig. 5.12

5.10 ENERGY STORED IN A CAPACITOR

When a p.d. is applied across a capacitor, the electrons are transferred from one plate to the other and the capacitor is charged. This involves expenditure of energy because electrons are moved against the opposing faces. This energy is stored in the electrostatic field set up between the plates of the capacitor in the dielectric medium. When the capacitor discharges, the field collapses and the stored energy is released.

Consider is capacitor of C farads connected across a source of V volt for charging. During charging, work is being done in shifting the charge (electrons) form one plate to other as shown in fig. 5.13. Let at any instant, the charge on the capacitor is q coulomb and p.d. across it is v volts.

Then, $C = q/v$

If further charged, the charge dq is shifted, then amount of work done is given by;

$$dw = v\,dq$$

or $$dw = C\,v\,dv \quad \begin{pmatrix} \because\ q = Cv \\ \therefore dq = C\,dv \end{pmatrix}$$

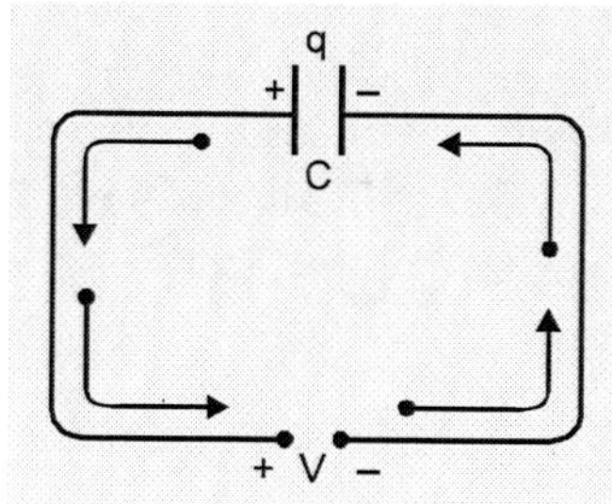

Fig. 5.13

∴ Total work done in raising the potential of uncharged capacitor 0 volt to V volt will be;

$$W = \int_0^V Cv\,dv = C\int_0^V v\,dv = C\left(\frac{v^2}{2}\right)_0^V = \frac{1}{2}CV^2 \text{ joule}$$

This work done or energy is stored in the electrostatic field set-up in the dielectric.

∴ Energy store in the capacitor or electrostatic field

$$W = \frac{1}{2}CV^2 = \frac{1}{2}QV = \frac{Q^2}{2C} \quad (\text{since } C = Q/V \text{ or } V = Q/C)$$

Example 5.1. *Five capacitor each of capacitance one micro-farad are joined in series, what will be the capacitance of the combination?*

Solution: Capacitance of each capacitor, $C = 1\ \mu$F

Let the total capacitance of combination be C_S, then

$$\frac{1}{C_S} = \frac{1}{C_1} + \frac{1}{C_2} + \ldots\ldots\ldots + \frac{1}{C_n} = \frac{1}{1} + \frac{1}{1} + \frac{1}{1} + \ldots\ldots 5 \text{ times}$$

$$= \frac{1+1+1+1+1}{1} = \frac{5}{1} \quad \therefore \quad C_S = \frac{1}{5}\mu F \ (Ans.)$$

Example 5.2. *Ten capacitors each of capacitance one micro-farad are joined in parallel, what will be the capacitance of the combination?*

Solution: Capacitance of each capacitor, $C = 1\ \mu$F

Let the total capacitance of the combination be C_p, then

$$C_p = C_1 + C_2 + C_3 + \ldots\ldots\ldots + C_n = 1 + 1 + 1 + \ldots\ldots\ldots\ldots \text{ten times}$$

$$C_p = \mathbf{10\ \mu F}\ (Ans.)$$

Example 5.3. *Capacitors of 4, 6 and 12* mF *are connected first in parallel and then in series. Compare the effective capacitance in the two combinations.*

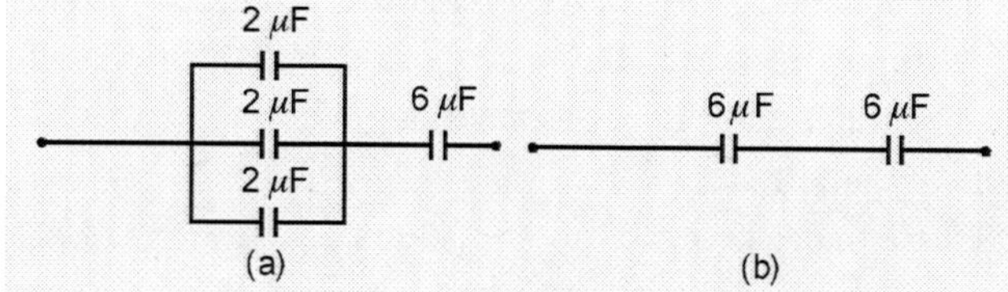

Fig. 5.14

Solution: When connected in parallel, let the capacitance be C_p

$$\therefore \quad C_p = 4 + 6 + 12 = 22\,\mu\text{F}$$

When connected in series, let the capacitance be C_S

$$\therefore \quad \frac{1}{C_S} = \frac{1}{4}+\frac{1}{6}+\frac{1}{12} = \frac{3+2+1}{12} = \frac{6}{12} = \frac{1}{2}$$

$$\therefore \quad C_S = 2\,\mu\text{F}$$

Ratio, $$\frac{C_p}{C_S} = \frac{22}{2} = \mathbf{11}\ (Ans.)$$

Example 5.4. *Three capacitors each of capacitance 2 mF are connected in parallel, this combination is connected in series with a fourth capacitor of capacitance 6 mF. Find the resultant capacitance of the combination.*

Solution: $C_p = C_1 + C_2 + C_3 = 2 + 2 + 2 = 6\,\mu\text{F}$

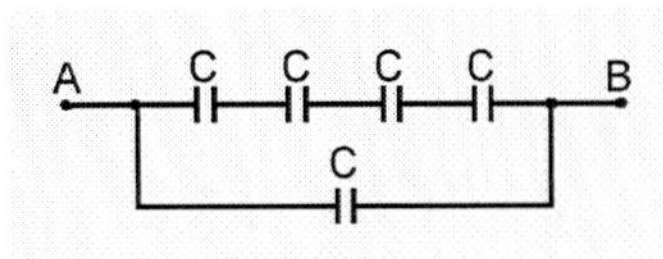

Fig. 5.15

$$\frac{1}{C_T} = \frac{1}{6}+\frac{1}{6} = \frac{1+1}{6} = \frac{1}{3}$$

$$C_T = 3\,\mu\text{F}$$

Example 5.5. *Four identical capacitors each of capacitance C farad are connected in series. One more similar capacitor is connected across this combination. Determine the capacitance of the whole combination.*

Solution: Let the capacitance of series combination be C_S, then

$$\frac{1}{C_S} = \frac{1}{C}+\frac{1}{C}+\frac{1}{C}+\frac{1}{C} = \frac{4}{C} \quad \text{or} \quad C_S = \frac{C}{4}$$

$$C_{AB} = C_S + C = \frac{C}{4} + C = \frac{1+4}{4} = \frac{5}{4}C = \mathbf{1.25\ C\ farad}\ (Ans.)$$

Example 5.6. *The capacitance of two capacitors is 25* mF *when connected in parallel and 6 mF when connected in series. Determine the capacitance of each capacitor.*

Solution: Let C_1 and C_2 be the capacitance of the two capacitors.

When connected in parallel, $$C_1 + C_2 = 25 \qquad ...(i)$$

When connected in series, $$\frac{C_1C_2}{C_1+C_2} = 6$$

or $\frac{C_1C_2}{25} = 6$ or $$C_1C_2 = 150 \qquad ...(ii)$$

Now, $C_1 - C_2 = \sqrt{(C_1+C_2)^2 - 4C_1C_2} = \sqrt{(25)^2 - 4\times150} = \sqrt{625-600}$

$$C_1 - C_2 = 5 \qquad ...(iii)$$

Adding eq. (*i*) and (*iii*), we get,

$$2C_1 = 30 \text{ or } C_1 = \mathbf{15\,\mu F} \text{ and } C_2 = \mathbf{10\,\mu F}\ (Ans.)$$

Example 5.6. *Two capacitors of capacity 2mF and 4 mF respectively are connected in series. A potential different of 900 V is applied between the extreme terminals. Find the potential difference across each capacitor.*

Solution: Capacitance of each capacitor, $C_1 = 2\ \mu F$

$C_2 = 4\ \mu F$

Voltage across the combination, $V = 900\ V$

$$\text{Total capacitance, } C = \frac{2 \times 4}{2+4} = \frac{4}{3}\mu F$$

$$\text{Charge on each capacitor, } Q = CV = \frac{4}{3} \times 900 = 1200\ \mu C$$

$$\text{P.d. across } 2\mu\text{F capacitor, } V_1 = \frac{Q}{C_1}$$

$$= \frac{1200 \times 10^{-6}}{2 \times 10^{-6}} = \textbf{600 V } (Ans.)$$

$$\text{P.d. across } 4\mu\text{F capacitor, } V_2 = \frac{Q}{C_2} = \frac{1200 \times 10^{-6}}{4 \times 10^{-6}} = \textbf{300 V } (Ans.)$$

Example 5.7. *Five capacitors each of capacitance C farad are connected as shown in fig. 5.16. Find the effective capacitance across AB.*

Solution: As the bridge is balance, the capacitor connected across branch *PQ* is neglected, then

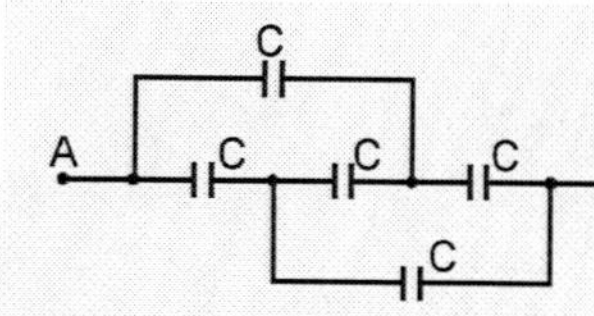

Fig. 5.16

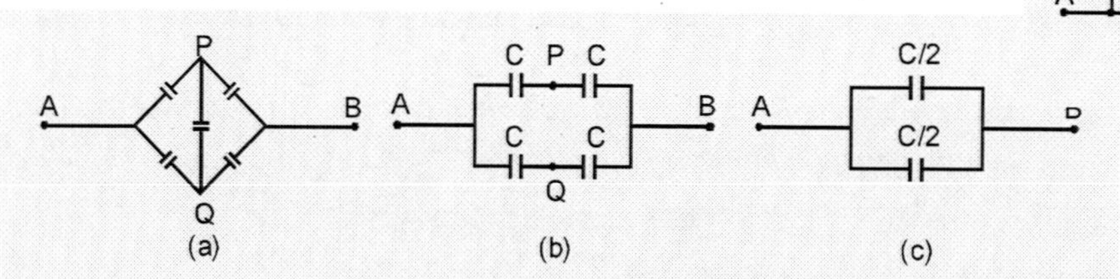

Fig. 5.17

$$\frac{1}{C_{APB}} = \frac{1}{C_{AQB}} = \frac{1}{C} + \frac{1}{C} = \frac{2}{C}$$

or
$$C_{APB} = C_{AQB} = \frac{C}{2}$$

$$C_{AB} = C_{APB} + C_{AQB} = \frac{C}{2} + \frac{C}{2} = \textbf{C farad } (Ans.)$$

Example 5.8. *Determine the capacitance of the condenser combination between terminals of fig. 5.18. The capacitances shown are in microfarads.*

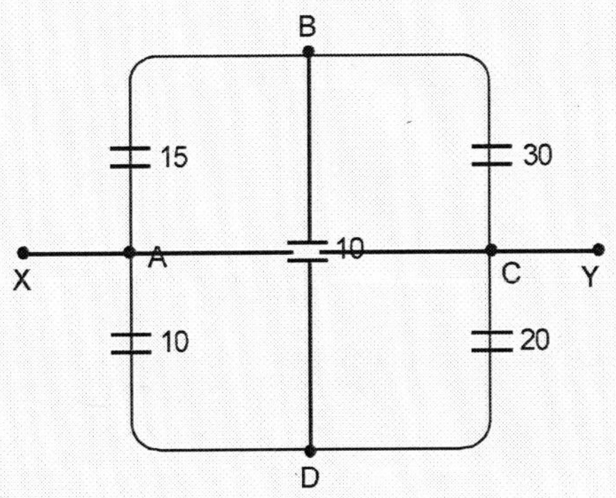

Fig. 5.18

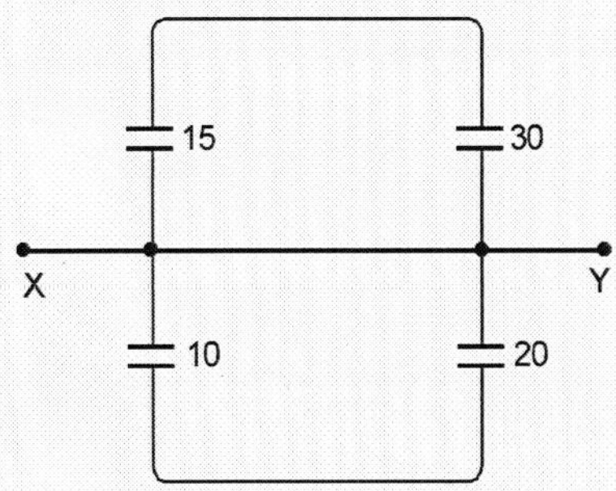

Fig. 5.19

Solution: Referring to fig. 5.19 the branches AB, BC, CD and DA are forming the four arms of a wheat stone bridge.

Now, $$\frac{C_{AB}}{C_{BC}} = \frac{15}{30} = 0.5$$

and $$\frac{C_{AD}}{C_{DC}} = \frac{10}{20} = 0 \cdot 5$$

Since the ratio arms $$\frac{C_{AB}}{C_{BC}} = \frac{C_{AD}}{C_{DC}}$$

The bridge is balanced and no current would flow through branch BD when some p.d. is applied across the terminals XY. Thus, branch BD can be removed and the circuit is reduced to as shown in fig. 5.19.

Now, in one branch 15μF and 30μF capacitors are connected in series with each other and in second branch 10μF and 20μF capacitors are connected in series with each other. The two branches are connected in parallel.

∴ The capacitance between the terminal X and Y.

$$C = \frac{15 \times 30}{15 + 30} + \frac{10 \times 20}{10 + 20}$$

$$= 10 + \frac{20}{3} = \frac{50}{3}$$

$$= \mathbf{16.67\ \mu\ F}\ (Ans.)$$

Example 5.9. *Two capacitors of capacitance 8mF and 2mF are connected in series across a 100 V d.c. supply. Now if the supply voltage is removed and the capacitors are then connected in parallel, what will be the final charge on each capacitor?*

Solution: Capacitance of first capacitor, $C_1 = 8\mu$F

Capacitance of second capacitor, $C_2 = 2\mu$F

When the two capacitors are connected in series across 100 V d.c. supply as shown in fig. 5.20 (a).

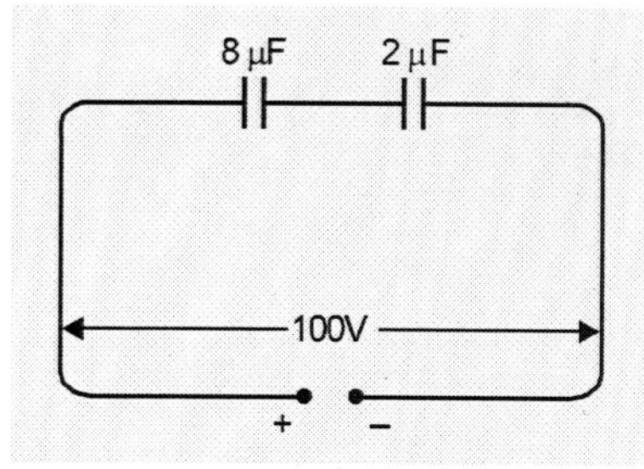

Fig. 5.20(a)

$$\text{Total capacitance, } C = \frac{C_1 \times C_2}{C_1 + C_2} = \frac{8 \times 2}{8 + 2} = 1 \cdot 6\ \mu\text{F}$$

$$\text{Total charge, } Q = CV$$

$$= 1.6 \times 100 = 160\ \mu\text{C}$$

Now the capacitors are connected in parallel after removing from the supply mains as shown in fig. 5.20(b). Let the charge on two capacitors be Q_1 and Q_2 respectively and the voltage across each of them is V volts.

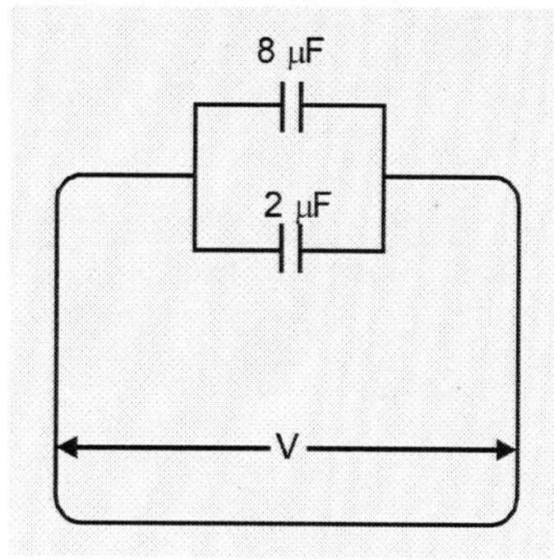

Fig. 5.20(b)

Total charge on the two capacitors remains the same *i.e.*

$$Q_1 + Q_2 = 160\ \mu C \quad \ldots(i)$$

$$\text{Voltage across each capacitor } V = \frac{Q_1}{C_1} = \frac{Q_2}{C_2}$$

or
$$\frac{Q_1}{8} = \frac{Q_2}{2}$$

or
$$Q_1 = 4\, Q_2 \quad \ldots(ii)$$

Substituting the value of Q_1 in equation (*i*), we get,

$$4\, Q_2 + Q_2 = 160$$

$$Q_2 = 32\ \mu\text{C}\ (Ans.)$$

and
$$Q_1 = 4 \times 32 = 128\ \mu\text{C}\ (Ans.)$$

TEST YOUR LEARNING-2

Type-I: Short Answer Questions

1. In parallel circuit, show that effective value of capacitance is equal to sum of the individual capacitances.
2. If n identical capacitors are connected in parallel and are charged to a potential of V volt. Now, they are separated and connected in series with additive polarities. What potential difference will be obtained across the combination.
3. Can we charge a capacitor beyond its capacity, justify.

Type-II: Numerical for Practice

1. Ten capacitors each of capacitance 1 μF are joined in (*i*) parallel, (*ii*) series. Find the resultant capacitance of each combination. (*Ans.* 10 μF, 0.1 μF)
2. Capacitors of 2, 3 and 6 μF capacitance are first connected in series and then in parallel, compare the effective capacitances of the two combinations. (*Ans.* $\frac{1}{11}$)

3. When two capacitors are connected in parallel their capacitance is 18 μF whereas their effective value becomes 4 μF when connected in series. Find the capacitance of each capacitor.

(*Ans.* 12 μF, 6 μF)

4. Five capacitors each of capacitance 2 μF are connected to form a wheat-stone bridge across terminal A and B. What will be the effective capacitance of the combination across AB. (*Ans.* 2C)
5. Three identical capacitors each of capacitance $3C$ farad are connected in series. This combination is connected in parallel with one more similar capacitor. Find the capacitance of whole combination. (*Ans.* 4C)

Type-III: Test Questions

1. Determine the value of capacitance of a grouping when (*i*) number of capacitors are connected parallel (*ii*) number of capacitors are connected in series.
2. Determine the effective value of capacitor's bank where three groups are connected in parallel and each group contains 2, 3 and 4 capacitors are connected in series.
3. Derive an expression for the energy stored in a capacitor. If the distance between the plates of a capacitor is decreased, whether its capacity to store energy will increase or decrease? Justify.

SELF ASSESMENT

TEST YOUR PREPARATION

OBJECTIVE TYPE QUESTIONS

Type-I: Fill in the blanks:

1. The two conducting surfaces separated by a dielectric medium is called a __________.
2. The charge on the capacitor plates per unit p.d. across the plate is called __________.
3. 1 pF capacitance = __________mF.
4. When a number of capacitors are connected in series, their effective value is __________ than the individual capacitance.

Type-II: Multi-choice Questions:

1. When two capacitors of capacitance C_1 and C_2 are connected in series, their effective value will be
   (*a*) more than C_1 and C_2 separately (*b*) less than C_1 and C_2 separately
   (*c*) equal to $C_1 + C_2$ (*d*) more than C_1 only
2. When two capacitors of 9 μF and 18 μF are connected in series, their effective value will be
   (*a*) 27 μF (*b*) 12 μF (*c*) 6μF (*d*) 21 μF
3. If two capacitors are charged and then connected in parallel after removing them from the batteries, their common potential is given by:
   (*a*) $V = V_1 + V_2$ (*b*) $V = V_1 - V_2$

(c) V = total charge/total capacitance (d) V = total capacitance/total charge

4. When four capacitors of capacitance 1/4 μF are connected in series, the resultant capacitance will be

(a) 1 μF (b) $\frac{1}{8}$ μF (c) $\frac{1}{16}$ μF (d) 4 μF

5. Three identical capacitors of capacitance C μF each are connected in series and such three combinations are connected in parallel. The capacitance of the whole combination will be

(a) C/3 μF (b) 3CμF (c) 1/3CμF (d) CμF

6. The number of 1μF capacitor needed to be connected in parallel in order to store a charge of 1 C with a p.d. of 500 V across the capacitors is

(a) 500 (b) 2000 (c) 4000 (d) none of these

7. Three identical capacitors of capacitance C each, are connected in series and this combination is connected in parallel with one more identical capacitor. The capacitance of the whole combination will be

(a) $C/3$ (b) $3/C$ (c) 4 $C/3$ (d) 3 $C/4$

VIVA-VOCE/REASONING QUESTIONS

1. A capacitor is also called a condenser, why?
2. A capacitor does not allow the electric current to pass through it, still it is used in the electric circuits, why?
3. The capacitance of a capacitor increases when air is replaced by mica between the plates of the capacitor, why?
4. When a charged capacitor is short circuited, a spark and noise is produced, why?
5. When a capacitor is connected in parallel with an existing capacitor, the the overall capacitance increases, why?

NUMERICALS FOR PRACTICE

1. Two capacitors having capacitances of 4μF and 6μF are connected in series across a 120 V d.c. supply. Calculate (i) the p.d. across each capacitor and (ii) the energy stored on each capacitor. (**Ans.** *72 V, 48 V, 0.010368 J, 0.0069 J*)

2. Two capacitors have capacitance of 6 μF and 10 μF respectively

   (i) Find the total capacitance when they are connected in

   (a) parallel (b) Series

   (ii) When the two capacitors are connected in series across a 200 V supply, find the p.d. across each capacitor and the charge on each capacitor.

   (**Ans.** *3.75 μF, 16 μF; 750 μC, 125 V, 75 V*)

3. Three capacitors of 10, 20 and 40μF are placed in series across a 350 V source. Determine:

   (i) Charge on each capacitor; (ii) Voltage drop across each capacitor;

   (**Ans.** 2×10^{-3} *C; 200 V, 100V, 50 V*)

4. Given some capacitor of 0.1 μF capable of withstanding up to 15 V each, calculate the number of capacitors needed if it is desired to obtain a capacitance of 0.1 μ F for use in circuit involving 60 V. (**Ans.** *16 i.e. 4 in series and 4 rows in parallel*)

5. Three capacitors A, B and C are connected in series across a 240 V d.c. supply. The p.d. across the three capacitors are 60 V, 80 V and 100 V respectively. If the capacitance of A is 12μF, What are capacitances of B and C ? (**Ans.** *9μF, 7.2 μF*)

6. When three capacitors A, B and C are connected in parallel their effective value is 60 μF and when they are connected in series their effective value is 60/11 μF. If the capacitance of A is 10 μ F, determine the capacitances of B and C. (**Ans.** *20 μF, 30 μF*)

SHORT ANSWER TYPE QUESTIONS

1. How will you define a capacitor.
2. How will you differentiate between capacitor and capacitance?
3. In series circuits, show that reciprocal of effective value of capacitance is equal to sum of the reciprocal of individual capacitances.
4. If n identical capacitors are connected in parallel and are charged to a potential of V volt. Now, if they are separated and connected in series with additive polarities. What potential difference will be obtained across the combination.

TEST QUESTIONS

1. Explain the construction of an electrolyte capacitor with the help of neat sketches. Why polarity is marked on the terminals of an electrolytic capacitor? If the terminals of this capacitor are connected in opposite polarity, what will happen?
2. Explain capacitor action. What are the factors on which the capacitance of a parallel plate capacitor depends?
3. Derive an expression for the equivalent capacitance for a number of capacitors connected in (*i*) Series (*ii*) Parallel.
4. Derive an expression for the potential energy stored in a charged condenser.
5. Derive an expression for the energy stored in an electric field.

ANSWERS TO OBJECTIVE TYPE QUESTIONS

(*i*) Fill in the blanks:

1. Capacitor **2.** Capacitance **3.** 10^{-6} **4.** less

(*ii*) Multi-Choice Questions:

1. (*b*) **2.** (*c*) **3.** (*c*) **4.**(*c*) **5.** (*d*)
6. (*b*) **7.** (*c*)

CHAPTER

6 ELECTROMAGNETIC EFFECTS

TOPICS TO BE DISCUSSED

Permanent magnets and Electromagnets, their construction and use, Polarities of an electromagnet and rules for finding them.

Faraday's Laws of Electromagnetic Induction, Dynamically induced e.m.f., its magnitude and induction, inductance and its unit. Mutually induced e.m.f., its magnitude and direction, Energy stored in an inductance.

Force acting on a current carrying conductor in magnetic field, its magnitude and direction, Principles and construction of dynamo.

LEARNING OBJECTIVES

*After the completion of this **chapter,** the students/readers will be able to understand:*

1. What are permanent and electro-magnets?
2. How to check the polarity of magnetic poles?
3. How to check the polarity of electro-magnets?
4. What is the meaning of electromagnetic induction?
5. What is dynamically and statically induces emf?
6. What is self and mutual inductance?
7. What happens when a current carrying conductor is placed in uniform magnetic field?
8. What are the main parts of a dynamo?

INTRODUCTION

The operation of all the electrical machines such as *d.c.* machines, transformers, synchronous machines, induction motors etc. rely upon their magnetic circuits. Electromagnetic effects play important role in the functioning of various electrical machines, equipment and devices.

In this chapter, we shall focus our attention on electromagnetic effects and their applications.

6.1 PERMANENT MAGNET AND MAGNETIC FIELD

A piece of iron that exibits magnetic field around it and has the tedancy to attract iron pieces is called a **magnet.** If it retains these properties permanently then it is called a **pemanent magnet.**

The region around the magnet where its poles exhibit a force of attraction or repulsion is called **magnetic field.**

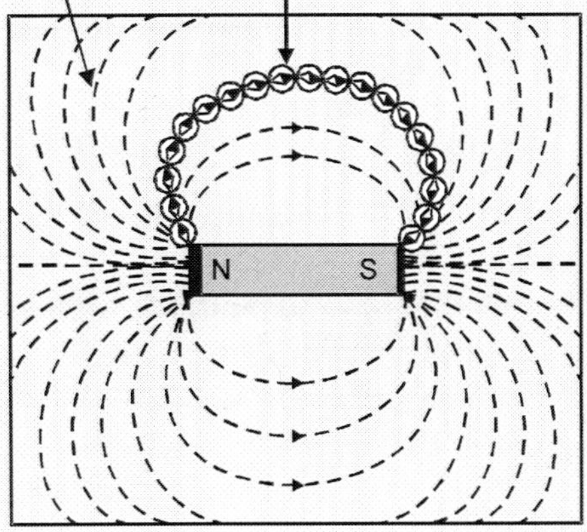

Fig. 6.1

The existence of the magnetic field at a point around the magnet can also be determined by placing a magnetic needle at that point as shown in fig. 6.1. Although magnetic lines of force have no real existence and are purely imaginary, yet their concept is very useful to understand various magnetic effects. It is assumed (because of their effects) that the magnetic lines of force possess following important properties:

(*i*) The direction of magnetic lines of force is from N-pole to the S-pole outside the magnet. But inside the magnet their direction is from S-pole to N-pole.

(*ii*) They form a closed loop.

(*iii*) Their tendency is to follow the least reluctance path.

(*iv*) They act like stretched cords, always trying to shorten themselves.

(*v*) They never intersect each other.

(*vi*) They repel each other when they are parallel and are in the same direction.

(*vii*) They remain unaffected by non-magnetic materials.

6.2 ELECTROMAGNETISM

The phenomenon by which magnetism is produced by the effect of electric current is called **electromagnetism.**

Magnetic effect of electric current

When a conductor carries electric currents, the magnetic lines of force are set up around it in the form of concentric circles. It can be illustrated by passing a current carrying conductor through a hard board as shown in fig. 6.2. Sprinkle iron filling on the hard board and tap it slightly, the iron filing will occupy a definite position around the conductor which reveals that:

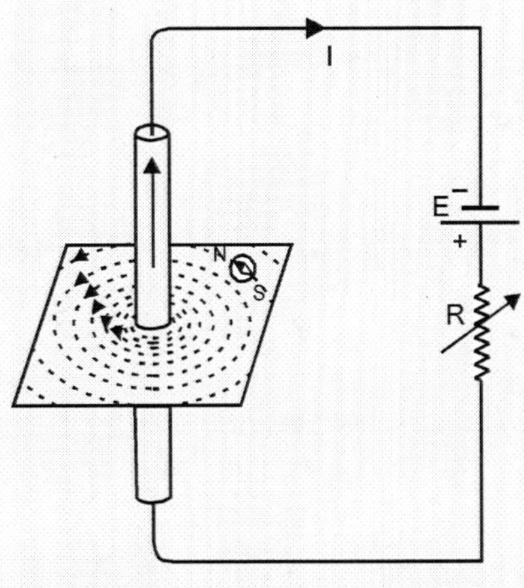

Fig. 6.2

(*i*) The field nearer the conductor is stronger and becomes weak as the radial distance from the conductor increases.

(*ii*) The strength of the magnetic field around the conductor depends upon the magnitude of flow of current through it. Larger the current, stronger is the filed and *vice-versa*.

(*iii*) The direction of magnetic field is perpendicular to direction of flow of current.

(*iv*) The direction of magnetic field is reversed if the direction of flow of current in the conductor is reversed. It can be detached by placing a magnetic needle near the conductor as shown in fig. 6.2.

6.3 DIRECTION OF MAGNETIC LINES OF FORCE

The direction of magnetic lines of force around a current carrying conductor is determined by applying;

(*i*) **Right hand thumb rule:** Hold the current carrying conductor in your right hand so that thumb points in the direction of flow of current [see fig. 6.3(*a*)]. The fingers curling around the conductor will point in the direction of magnetic lines of force around it.

(*ii*) **Right hand cork screw rule:** Hold the right handed screw on the cork and rotate if such that it advances in the direction of flow of current [see fig. 6.3(*c*)]. The direction of rotation of the screw gives us the direction of magnetic lines of force around the current carrying conductor.

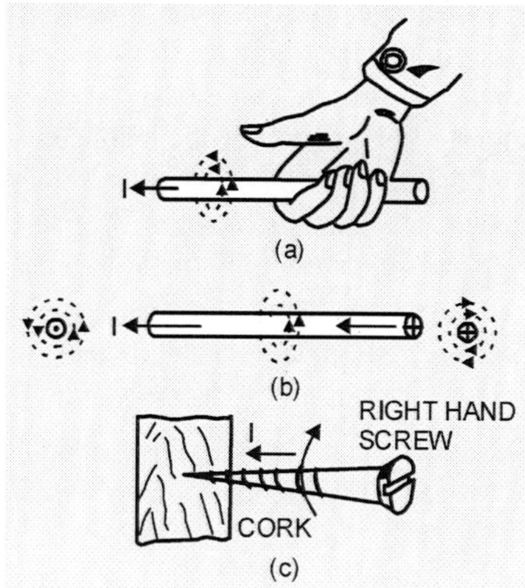

Fig. 6.3

Hence, it is concluded that when the flow of current is out of the plane of the conductor (represented by dot in a circle), the magnetic lines of force are set-up in anticlockwise direction, as shown in fig. 6.3 (*b*). Whereas, when the flow of current is into the plane of the conductor (represented by cross in a circle), the magnetic lines of force are set-up in clockwise direction.

6.4 TYPICAL ELECTROMAGNETIC FIELDS

Different electromagnetic fields are produced by the current carrying conductors wound in different shapes. Some of them are illustrated below:

(*i*) **Magnetic field produced by a circular coil:** The magnetic field set by a circular coil is shown in fig. 6.4. The magnetic field sets-up a polarity on the two sides of the coil which can be determined with the help of magnetic needle placing on either side of the coil as shown in fig. 6.4(*b*).

(*ii*) **Magnetic field produced by a solenoid:** A number of circular coils connected in series carrying the same current in same direction and arranged co-axially is called a *solenoid*. In fact, solenoid is a Greek word, meaning tube-like. The solenoid may be iron cored on air cored, as shown in fig. 6.5. When current flows through the solenoid magnetic field is set-up around it. The polarity of the solenoid is determined by *Righ hand thumb rule*, stated below:

Hold the solenoid in your right hand so that the fingers curling around point in the direction of flow of current in the coils, then thumb stretched along the axis of the solenoid will point towards the N-pole as shown in fig. 6.6.

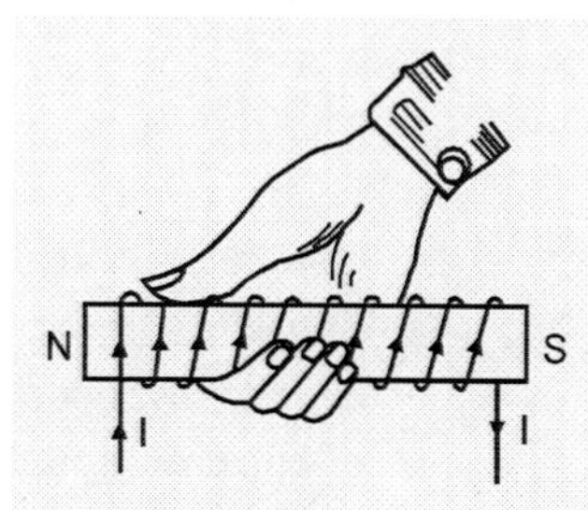

Fig. 6.6

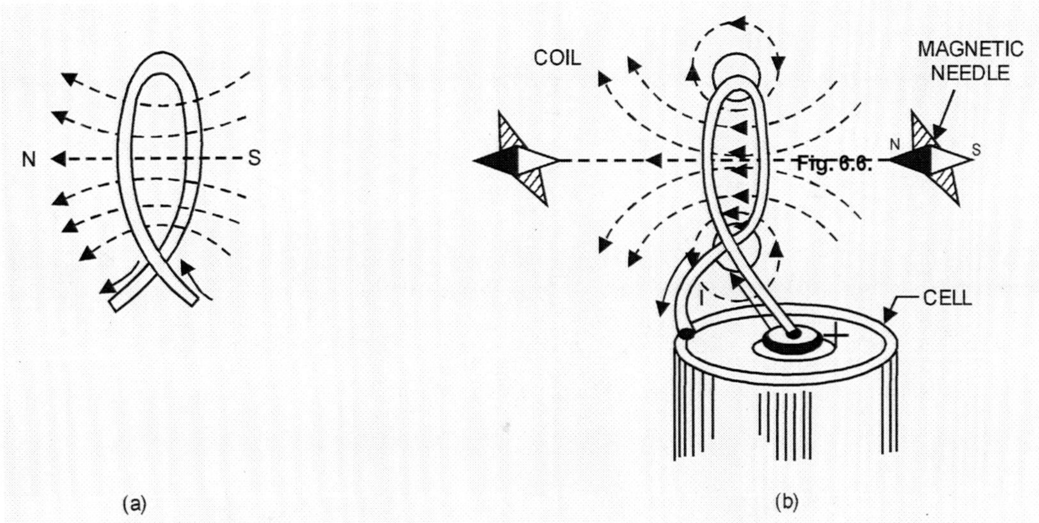

Fig. 6.4

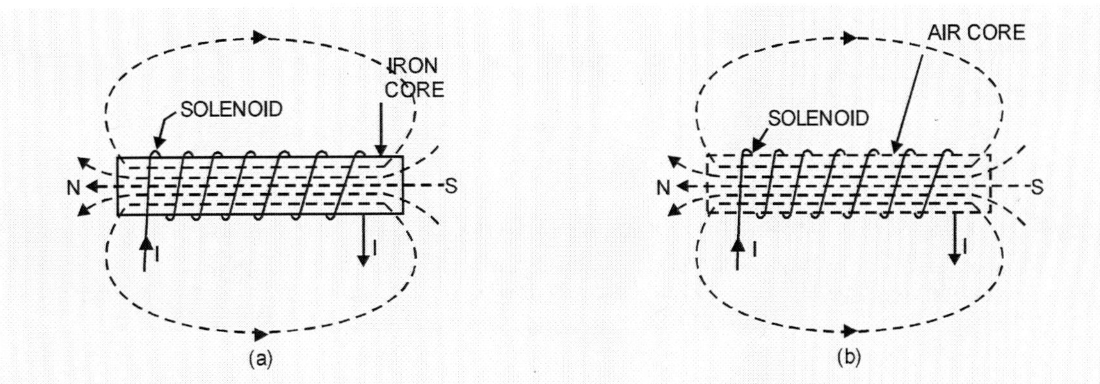

Fig. 6.5

6.5 ELECTROMAGNET

A magnetic material when obtains the magnetic properties by providing a current carrying solenoid around it is called an **electromagnet.**

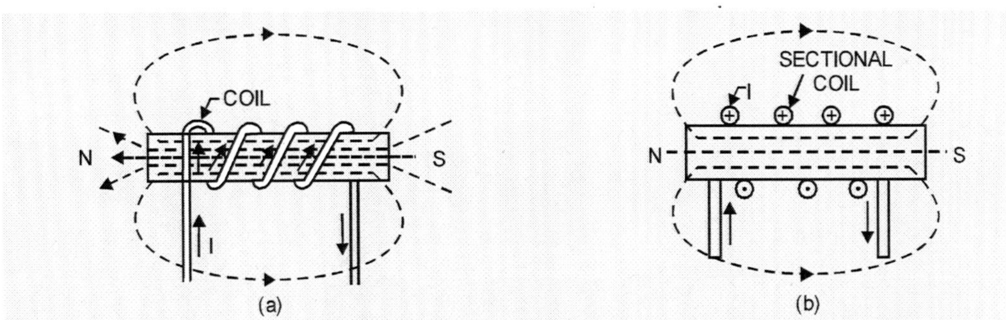

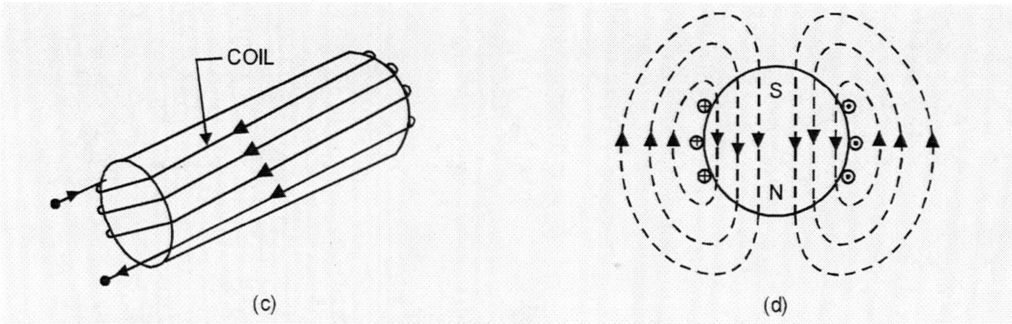

Fig. 6.7

An electromagnet may have different shapes as shown in fig. 6.7.

Electromagnets are employed in all electrical machines, relays, magnetic jaws etc.

TEST YOUR LEARNING - 1

Type-I: Short Answer Questions

1. What do you mean by permanent magnets and electromagnets?
2. Define magnetic field.
3. How will you determine the polarity of poles of a permanent magnet?
4. How will you determine the polarity of poles of an electromagnet?

Type-II: Test Questions

1. How will you apply Right Hand Thumb Rule to determine the polarity of poles in an electromagnet?
2. What are the applications of electromagnets?
3. How the direction of magnetic lines of force is determined set-up around a current carrying conductor?

6.6 ELECTRO-MAGNETIC INDUCTION

The phenomenon by which an e.m.f. is induced in a circuit (an hence current flows when the circuit is closed) when magnetic flux linking with it changes is called *electro-magnetic induction.*

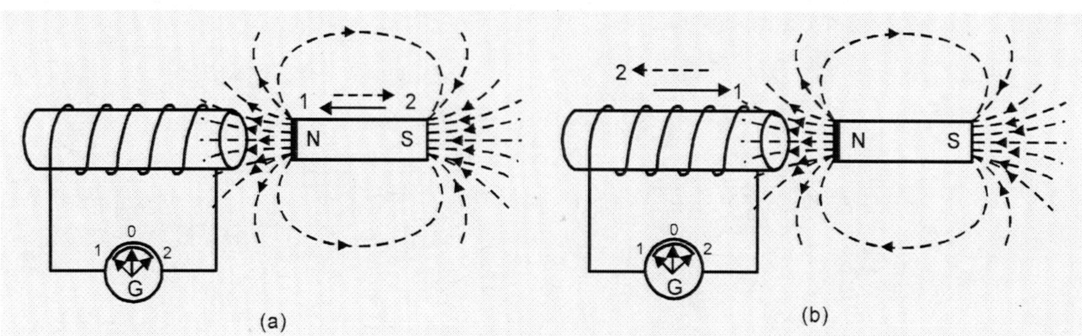

Fig. 6.8

For illustration, consider a coil having a large number of turns to which galvanometer is connected. When a permanent bar magnet is taken nearer to the coil or away from the coil, as shown in fig. 6.8 (*a*), a deflection occurs in the needle of the galvanometer. Although, the deflection in the needle is opposite is two cases.

On the other hand, if the bar magnet is kept stationary and the coil is brought nearer to the magnet or away from the magnet, as shown in fig. 6.8 (*b*), again a deflection occurs in the needle of the galvanometer. The deflection in the needle is opposite in the two cases.

However, if the magnet and the coil both are kept stationary, no matter how much flux is linking with the coil, there is no deflection in the galvanometer needle.

The following points are worth noting:

(*i*) The deflection in he galvanometer needle shows that e.m.f. is induced in the coil. This condition occurs only when flux linking with the circuit changes *i.e.* either magnet or coil is in motion.

(*ii*) The direction of induced e.m.f. in the coil depends upon the direction of magnetic field and the direction of motion of coil.

6.7 FARADAY'S LAWS OF ELECTROMAGNETIC INDUCTION

Michael Faraday summed up conclusions of his experiments regarding electro-magnetic induction into two laws, known as *Faraday's laws of electro-magnetic induction*.

First Law: This law states that "*Whenever a conductor cuts across the magnetic field, an e.m.f. is induced in the conductor.*"

or

"*Whenever the magnetic flux linking with any circuit (or coil) changes, an e.m.f. is induced in the circuit.*"

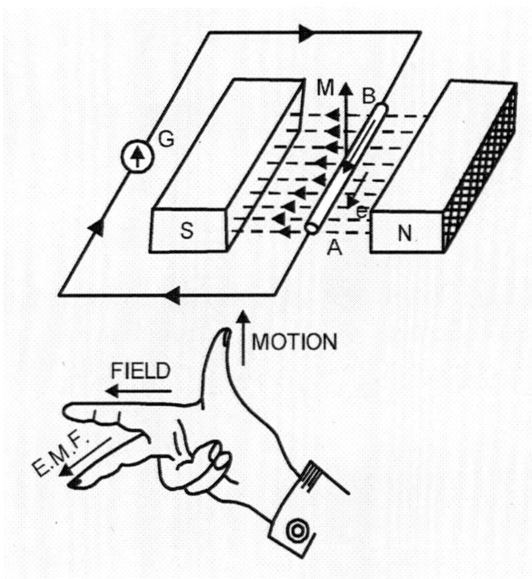

Fig. 6.9

Fig. 6.9 shows a conductor placed in the magnetic field of a permanent magnet to which a galvanometer is connected. Whenever, the conductor is moved upward or downward *i.e.* across the field, there is deflection in the galvanometer needle which indicates that an e.m.f. is induced in the conductor. If the conductor is moved along (parallel) the field, there is no deflection in the needle which indicates that no e.m.f. is induced in the conductor.

For the second statement, consider a coil placed near a bar magnet and a galvanometer connected across the coil, as shown in fig. 6.10. When the bar magnet (N-pole) is taken nearer to the coil [see fig. 6.10 (*a*)], there is deflection in the needle of the galvanometer. If now the bar magnet (*N*-pole) is taken away from the coil [see fig. 6.10 (*b*)], again there is deflection in the needle of galvanometer but in opposite direction. The deflection in the needle of galvanometer indicates that e.m.f. is induced in the coil.

Second Law: This law states that "*The magnitude of induced e.m.f. in a coil is directly proportional to the rate of change of flux linkages*.

Rate of change of flux linkages $= \dfrac{N(\phi_2 - \phi_1)}{t}$ Wb-turns/s

where, N = No. of turns of the coil;

$(\phi_2 - \phi_1)$ = change of flux in Wb

t = time in second for the change

∴ According to Faraday's second law of electro-magnetic induction;

Induced e.m.f., $$e \propto \frac{N(\phi_2 - \phi_1)}{t}$$

$$e = \frac{N(\phi_2 - \phi_1)}{t} \text{ (taking proportionality constant, as unity)}$$

In differential form, $$e = N\frac{d\phi}{dt} \text{ volt}$$

Usually, a minus sign is given to the right-hand side expression which indicates that e.m.f. is induced in such a direction which opposes the cause (*i.e.* change in flux) that produces it (according to Lenz's law).

$$e = -N\frac{d\phi}{dt} \text{ volt}$$

6.8 DIRECTION OF INDUCED E.M.F.

The direction of induced e.m.f. and hence current in a conductor or coil can be determined by either of the following two methods:

1. Flaming's Right hand Rule: This rule is applied to determine the direction of induced e.m.f. in a conductor moving across the field and is stated as under;

"Stretch, first finger, second finger, and thumb of your right hand mutually perpendicular to each other. If first finger indicates the direction of magnetic field, thumb indicates the direction of motion of conductor then second finger will indicate the direction of induced e.m.f. in the conductor."

Its illustration is shown in fig. 6.9.

2. Lenz's Law: This law is more suitably applied to determine the direction of induced e.m.f. in a coil or circuit when flux linking with it changes. It is stated as under:

"In effect, electro-magnetically induced e.m.f. and hence current flows in a coil or circuit in such a direction that the magnetic field set up by it, always opposes the very cause which produces it."

Explanation: When N-pole of a bar magnet is taken nearer to the coil as shown in fig. 6.10 (*a*), an e.m.f. is induced in the coil and hence current flows through it in such a direction that side 'B' of the coil attains North polarity which opposes the movement of the bar magnet.

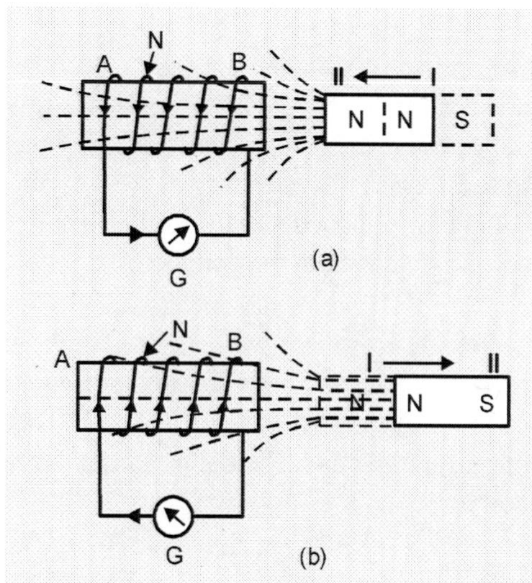

Fig. 6.10

Whereas, when N-pole of the bar magnet is taken away from the coil as shown in fig. 6.10 (*b*), the direction of e.m.f. induced in the coil is reversed and side 'B' of the coil attains South polarity which again opposes the movement of the bar magnet.

6.9 INDUCED E.M.F.

When flux linking with a conductor (or coil) changes, an e.m.f. is induced in it. This change in flux linkages can be obtained in the following two ways:

(*i*) By either moving the conductor and keeping the magnetic field system stationary or moving the magnetic field system and keeping the conductor stationary, in such a way that conductor cuts across the magnetic field (as in case of d.c. and a.c. generators). The e.m.f. induced in this way is called *dynamically induced e.m.f.*

(*ii*) By changing the flux linking with the coil (or conductor) without moving either coil or field system. However, the change of flux produced by the field system linking with the coil is obtained by changing the current in the field system (solenoid), as in transformers. The e.m.f. induced in this way is called *statically induced e.m.f.*

6.10 DYNAMICALLY INDUCED E.M.F.

By either moving the conductor keeping the magnetic field system stationary or moving the field system keeping the conductor stationary so that flux is cut by the conductor, the e.m.f. thus induced in the conductor is called *dynamically induced e.m.f.*

Mathematical expression

Considering a conductor of length l metre placed in the magnetic field of flux density B Wb/m^2 is moving at right angle to the field at a velocity u metre/second as shown in fig. 6.11 (*a*). Let the conductor be moved through a small distance dx metre in time dt second as shown in fig. 6.11 (*b*).

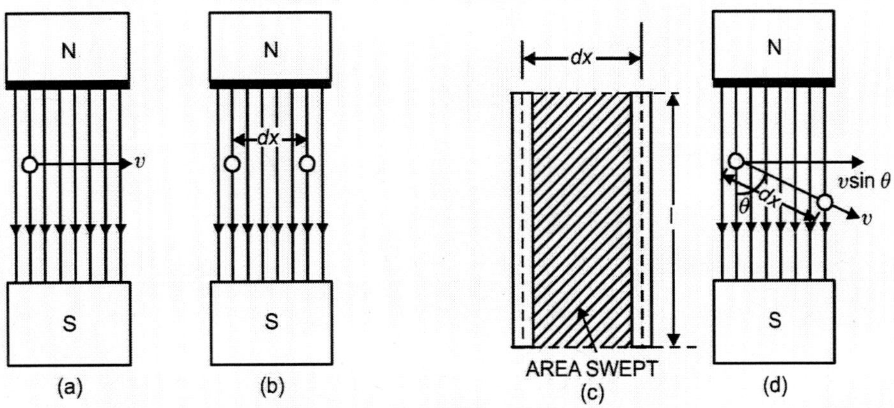

Fig. 6.11

Area swept by the conductor, $A = l \times dx$

Flux cut by the conductor, $\phi = B \times A = B\, l\, dx$

According to Faraday's Law of electro-magnetic induction;

Induced e.m.f., $e = \frac{\text{flux cut}}{\text{time}} = \frac{\phi}{dt} = \frac{B\,l\,dx}{dt} = B\,l\,\upsilon$ (since $dx/dt = \upsilon$ (velocity))

Now, if the conductor is moved at an angle θ with the direction of magnetic field at a velocity υ metre/second as shown in fig. 29 (*d*). A small distance covered by the conductor in that direction is *dx* in time *dt* second. Then the component of distance perpendicular to the magnetic field, which produces e.m.f., is $dx \sin \theta$.

$\therefore$ Area swept by the conductor, $A = l \times dx \sin \theta$

Flux cut by the conductor, $\phi = B \times A = B\,l\,dx \sin \theta$

$$\text{Induced e.m.f. } e = \frac{B\,l\,dx \sin \theta}{dt} = B\,l\,\upsilon \sin \theta$$

Example 6.1. *A coil of 500 turns in linked with a flux of 2 mWb. If this flux is reversed in 4 ms, calculate the average e.m.f. induced in the coil.*

Solution: Average induced e.m.f., $e = N\frac{d\phi}{dt}$

where, $N = 500$ turns; $d\phi = 2 - (-2) = 4\ m\ Wb$; $dt = 4 \times 10^3\ s$;

$\therefore$ $e = 500 \times \frac{4 \times 10^{-3}}{4 \times 10^{-3}} =$ **500 V** (*Ans.*)

Example 6.2. *A coil of 250 turns is wound on a magnetic circuit of reluctance 100000 AT/Wb. If a current of 2 A flowing in the coil is reversed in 5 ms, find the average e.m.f. induced in the coil.*

Solution: ϕ = m.m.f./reluctance *i.e.* $\phi = NI/S$

where, $N = 250$; $I = 2$A and $S = 100000$ AT/Wb.

$\therefore$ $\phi = \frac{250 \times 2}{100000} = 5$ m Wb

Average induced e.m.f. $e = N\frac{d\phi}{dt}$

where, $d\phi = 5 - (-5) = 10$ m Wb (since current is reveresed)

$e = 250 \times \frac{10 \times 10^{-3}}{5 \times 10^{-3}} =$ **500 V** (*Ans.*)

6.11 STATICALLY INDUCED E.M.F.

When the coil and magnetic field system both are stationary but the magnetic field linking with the coil changes (by changing the current producing the field), the e.m.f. thus induced in the coil is called *statically induced e.m.f.*

The statically induced e.m.f. may be:

(*i*) Self induced e.m.f. (*ii*) Mutually induced e.m.f.

(*i*) Self induced e.m.f.: The e.m.f. induced in a coil due to the change of flux produced by it linking with its own turns, as shown in fig. 6.12, is called *self induced e.m.f.*

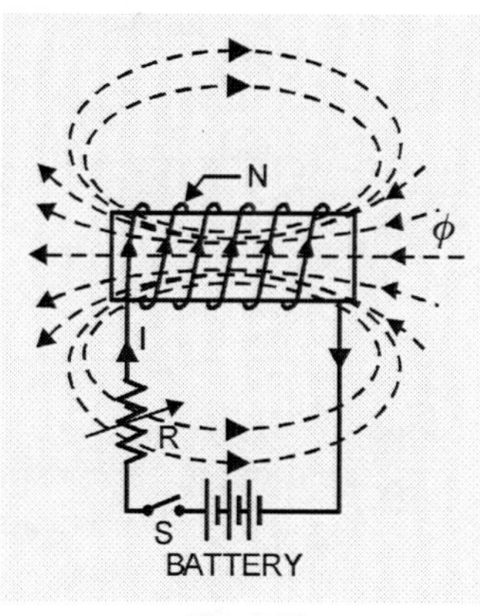

Fig. 6.12

The direction of this induced e.m.f. is such that it opposes the cause which produces it (lenz's law) *i.e.* change of current in the coil.

Since the rate of change of flux linking with the coil depends upon the rate of change of current in the coil. Therefore, the magnitude of self induced e.m.f. is directly proportional to the rate of change of current in the coil. Therefore, the magnitude of self induced e.m.f. is directly proportional to the rate of change of current in the coil, *i.e.*

$$e \propto \frac{dI}{dt} \text{ or } e = L\frac{dI}{dt}$$

Where L is a constant of proportionality and is called *self inductance or co-efficient of self inductance or inductance of the coil.*

***(ii)* Mutually induced e.m.f.**

The e.m.f. induced in a coil due to the change of flux produced by another (neighbouring) coil, linking with it is called *mutually induced e.m.f.*

Since the rate of change of flux linking with coil 'B', as shown in fig. 6.13, depends upon the rate of change of current in coil 'A'. Therefore, the magnitude of mutually induced e.m.f. is directly proportional to the rate of change of current in coil 'A', *i.e.*

$$e_m \propto \frac{dI_1}{dt} \text{ or } e_m = M\frac{dI_1}{dt}$$

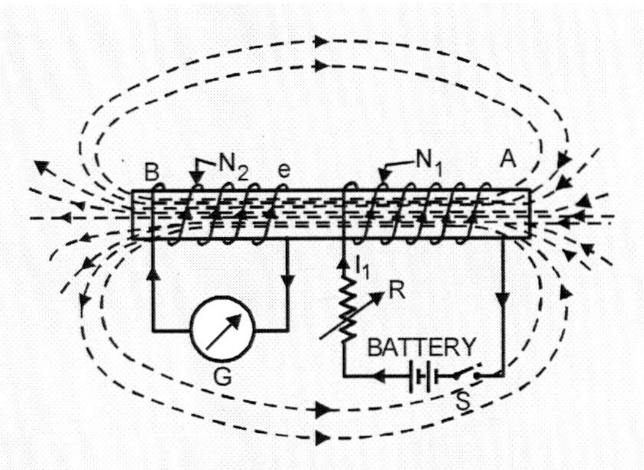

Fig. 6.13

Where M is a constant of proportionality and is called *mutual inductance* or *co-efficient of mutual inductance.*

6.12 SELF INDUCTANCE

The property of a coil due to which it opposes the change of current flowing through itself is called **self inductance** *or* **inductance** *of the coil.*

This property (*i.e.* inductance) is attained by a coil due to self-induced e.m.f. produced in the coil itself by the changing current flowing through it. If the current in the coil is increasing (by the change in circuit conditions), the self-induced e.m.f. is produced in the coil in such a direction so as to oppose the rise of current *i.e.* the direction of self-induced e.m.f. is opposite to that of the applied voltage. On the other hand, if the current in the coil is decreasing, the self-induced e.m.f. is produced in the coil in such direction so as to oppose the fall of current *i.e.* the direction of self-induced e.m.f. is in the same direction as that of the applied voltage. In fact, self-inductance does not prevent the change of current but it delays the change of current flowing through a coil.

It may be noted that this property of the coil only opposes the changing current (*i.e.* alternating current). However, it does not effect the steady (*i.e.* direct) current when flows through it. In other words, the self-inductance of the coil (by virtue of its geometrical and magnetic properties) will exhibit its presence to the alternating current but it will not exhibit its presence to the direct current.

Expressions for self inductance:

$$L = \frac{e}{dI/dt} \quad \left(\text{since } e = L\frac{dI}{dt}\right)$$

$$= \frac{N\phi}{I} \left(\text{since } e = N\frac{d\phi}{dt} = L\frac{dI}{dt}\right) = \frac{N^2}{l/a\mu_0\mu_r} \left(\text{since } \phi = \frac{NI}{l/a\mu_0\mu_r}\right)$$

6.13 MUTUAL INDUCTANCE

*The property of one coil due to which it opposes the change of current in the other (neighbouring) coil is called **mutual-inductance** between the two coils.*

This property (*i.e.* mutual-inductance) is attained by a coil due to mutually induced e.m.f. in the coil while current in the neighbouring coil is changing.

Expression for mutual inductance:

$$M = \frac{e_m}{dI_1/dt} \quad \left(\text{since } e_m = M\frac{dI_1}{dt}\right)$$

$$= \frac{N_2\phi_{12}}{I_1} \left(\text{since } e_m = N_2\frac{d\phi_{12}}{dt} = M\frac{dI_1}{dt}\right) = \frac{N_1N_2}{l/a\mu_0\mu_r} \left(\text{since } \phi_{12} = \frac{N_1I_1}{l/a\,\mu_0\mu_r}\right)$$

Example 6.3. *A coil has 1500 turns. A current of 4 A causes a flux of 8 mWb to link the coil. Find the self inductance of the coil.*

Solution: Inductance of the coil, $L = \frac{N\phi}{I}$

where $N = 1500$; $\phi = 8 \times 10^{-3}$ Wb and $I = 4$ A.

$\therefore$ $L = \frac{1500 \times 8 \times 10^{-3}}{4} = 3$ **H** (*Ans.*)

Example 6.4. *Calculate the value of e.m.f. induced in circuit having an inductance of 700 micro-henry if the current flowing through it varies at a rate of 5000 A per second.*

Solution: Inductance of the coil, $L = 700 \times 10^{-6}$ H

Rate of change of current, $\frac{dI}{dt} = 5000$ A/s

$\therefore$ Magnitude of e.m.f. induced in the coil,

$$e = L\frac{dI}{dt} = 700 \times 10^{-6} \times 5000 = 3.5 \text{ V } (Ans.)$$

Example 6.5. *An air cored solenoid has 300 turns its length is 25 cm and its cross section is 3 cm². Calculate the self-inductance is henry.*

Solution: No. of turns of the solenoid, $N = 300$

Length of solenoid, $l = 25$ cm $= 0.25$ m
Area of cross section, $a = 3$ cm$^2 = 3 \times 10^{-4}$ m^2
For air core, $\mu_r = 1$

Inductance of the solenoid, $$L = \frac{N^2}{l/a\,\mu_0\,\mu_r} = \frac{N^2}{l} a\,\mu_0$$

$$= \frac{300 \times 300}{0 \cdot 25} \times 3 \times 10^{-4} \times 4\pi \times 10^{-7} = \mathbf{0.1375\ mH}\ (Ans.)$$

Example 6.6. *Calculate the inductance of toroid, 25 cm mean diameter and 6.25 cm² circular cross-section wound uniformly with 1000 turns of wire. Hence calculate the e.m.f. induced when current in it increases at the rate of 100 A/second.*

Solution: Inductance of the toroid, $L = \frac{N^2}{l} \times a\,\mu_0\,\mu_r$

where, No. of turns, $N = 1000$ turns
Mean length $l = \pi D = 0.25\,\pi$ m;
Area of x-section, $a = 6.25 \times 10^{-4}$ m^2 and
Relative permeability, $\mu_r = 1$

$\therefore$ $L = (1000)^2 \times 6.25 \times 10^{-4} \times 4\,\pi \times 10^{-7} \times 1/0.25\,\pi = \mathbf{1\ m\ H}$ (*Ans.*)

Induced e.m.f., $e = L\frac{dI}{dt} = 1 \times 10^{-3} \times 100 = \mathbf{0.1\ V}$ (*Ans.*)

Example 6.7. *The iron core of a choke has mean length 25 cm with an air gap of 1 mm. The choke is designed for an inductance of 15 H when operating at a flux density of 1 Wb/m². The iron core has a relative permeability of 3000 and 8 cm² area of cross-section. Determine the required number of turns of the coil.*

Solution: Inductance of the coil, $L = N^2/S_T$
where S_T is the total reluctance of the magnetic circuit

$$S_T = \frac{l_i}{a\,\mu_0\,\mu_r} + \frac{l_g}{a\,\mu_0}$$

$$= \frac{0 \cdot 25}{8 \times 10^{-4} \times 4\pi \times 10^{-7} \times 3000} + \frac{1 \times 10^{-3}}{8 \times 10^{-4} \times 4\,\pi \times 10^{-7}} = 1077612 \text{ AT/Wb}$$

Now $$N = \sqrt{L\,S_T} = \sqrt{15 \times 1077612} = \mathbf{4020.5\ turns}\ (Ans.)$$

Example 6.8. *Two coils have a mutual inductance of 0.6 H. If current in one coil is varied from 4 A to 1 A in 0.2 second, calculate (i) the average e.m.f. induced in the other coil and (ii) the change of flux linking the later assuming that it is wound with 150 turns.*

Solution: Mutually induced e.m.f., $e_m = M\frac{dI_1}{dt}$

where, $M = 0.6$ H; $dI_1 = 4 - 1 = 3$ A and $dt = 0.2$ s

$\therefore$ $e_m = 0.6 \times 3/0.2 = \mathbf{9\ V}$ (*Ans.*)

Now, $$e_m = N_2 \frac{d\phi_{12}}{dt}$$

∴ Change of flux with second coil,

$$d\phi_{12} = \frac{e_m \times dt}{N_2} = \frac{9 \times 0 \cdot 2}{150} = \textbf{12 mWb} \ (Ans.)$$

6.14 ENERGY STORED IN A MAGNETIC FILED

When some electrical energy is supplied to a coil, it is spent in two ways:

(*i*) A part of it is spent to meet $I^2 R$ loss which is dissipated in the form of heat and cannot be recovered.

(*ii*) The remaining part is used to create magnetic field around the coil and is stored in the magnetic field. When this field collapses, the stored energy is released by the coil and is returned to the circuit.

The energy stored in the magnetic field is given by the expression:

$$\text{Energy stored in magnetic field} = \frac{1}{2} L I^2$$

Example 6.9. *A solenoid of 1 m in length and 10 cm in diameter has 5000 turns. Calculate the energy in the magnetic field when a current of 2 A flows in the solenoid.*

Solution: Inductance of the solenoid, $L = \frac{N^2}{l} \times a \mu_0 \mu_r$

where, $N = 5000$; $a = \pi d^2/4 = 25\,\pi \times 10^{-4}$ m²; $l = 1$ m; $\mu_r = 1$.

∴ $L = (5000)^2 \times 25\,\pi \times 10^{-4} \times 4\,\pi \times 10^{-7} \times 1/1 = 0.2467$ H

$$\text{Energy stored} = \frac{1}{2} LI^2 = \frac{1}{2} \times 0 \cdot 2467 \times (2)^2 = \textbf{0.4934 J} \ (Ans.)$$

TEST YOUR LEARNING-2

Type-I: Short Answer Questions (*Section–A: 2 marks each*)

1. What do mean by electro-magnetic induction?
2. Define Faraday's laws of electromagnetic induction.
3. State Fleming's Right Hand Rule.
4. State Lenz's law.
5. What do you mean by dynamically induced emf?
6. What do you mean by statically induced emf? How self induced emf is different to mutually induced emf?
7. What do you understand by self inductance?
8. What is mutual inductance?
9. How can you determine the energy stored in a magnetic field?

Type-II: Numerical for Practice

1. A coil of 500 turns in wound on a magnetic circuit of reluctance 2×10^5 AT/Wb. If a current of 2 A flowing through the coil is reversed in 10 ms, find the average e.m.f. induced in the coil.
   (*Ans. 300 V*)
2. An iron ring or toroid 0.2 m in diameter and 10 sq. cm cross-sectional area of the core, is wound with 250 turns of wire. If the flux density in the core is to be 1 Wb/m^2 and relative permeability of iron is 500, what is the exciting current required to be passed in the winding? Determine also the self inductance. (*Ans. 4 A; 62.5 m H*)
3. Two identical coils A and B each having 1500 turns lie in parallel planes such that 60% of the flux produced by one coil links with the other. A current of 10 A in coil A produces a flux of 1 m Wb in it. If the current in coil A changes from + 15 A to – 15 A in 0.02 second, (*i*) What would be the mutual inductance and magnitude of the e.m.f. induced in coil B? (*ii*) Calculate the self inductance of each coil. (*Ans. 0.09 H, 135 V; 0.15 H*)
4. Two coils *A* and *B*, have self inductance of 240 μH and 600 μH respectively. When a current of 3 A through coil A is reversed, the deflection on the flux meter connected across *B* is 1200 μ Wb-turns. Calculate (*i*) the mutual inductance between the coils; (*ii*) the average e.m.f. induced in coil *B* if the flux is reversed in 0.2 second and (*iii*) co-efficient of coupling.
   (*Ans.* 200 μ *H; 6 m V; 0.527*)

Type-III: Test Questions (*Section: B & C – 5 marks each*)

1. State and explain Faraday's laws of electro-magnetic induction.
2. State and explain Fleming's Right Hand Rule and Lenz's law. How these are useful to determine the direction of electromagnetically induced emf.
3. Show that dynamically induced emf $e = Blv \sin \theta$.
4. Define and explain statically induced emf. How self induced emf is different to mutually induced emf?
5. Define and explain self and mutual inductance.
6. Show that energy stored in a magnetic field is $= \frac{1}{2} LI^2$

6.15 CURRENT CARRYING CONDUCTOR IN MAGNETIC FIELD

When a current carrying conductor is placed in a uniform magnetic field perpendicular to it, a *mechanical force* is experienced by the conductor.

To understand this phenomenon, consider a uniform magnetic field set-up between two opposite poles as shown in fig. 6.14 (*a*). The magnetic field produced by the current carrying conductor is shown in fig. 6.14 (*b*). When this current carrying conductor is placed in the uniform magnetic field, a resultant field is set-up by the interaction of the two fields as shown in fig. 6.14 (*c*). This resultant field behaves like a stretched rubber band the exerts a force perpendicular to both the main magnetic field and the direction of flow of current, as shown in fig. 6.14 (*c*).

The direction of force acting on the conductor is determined by *Flaming's Left Hand Rule stated* below:

Stretch first finger, second finger and thumb of your left hand mutually perpendicular to each other [see fig. 6.14 (*c*)]. If first finger represents he direction of main field, second finger represents the direction of flow of current through the conductor, them thumb will indicate the direction of force acting on the conductor.

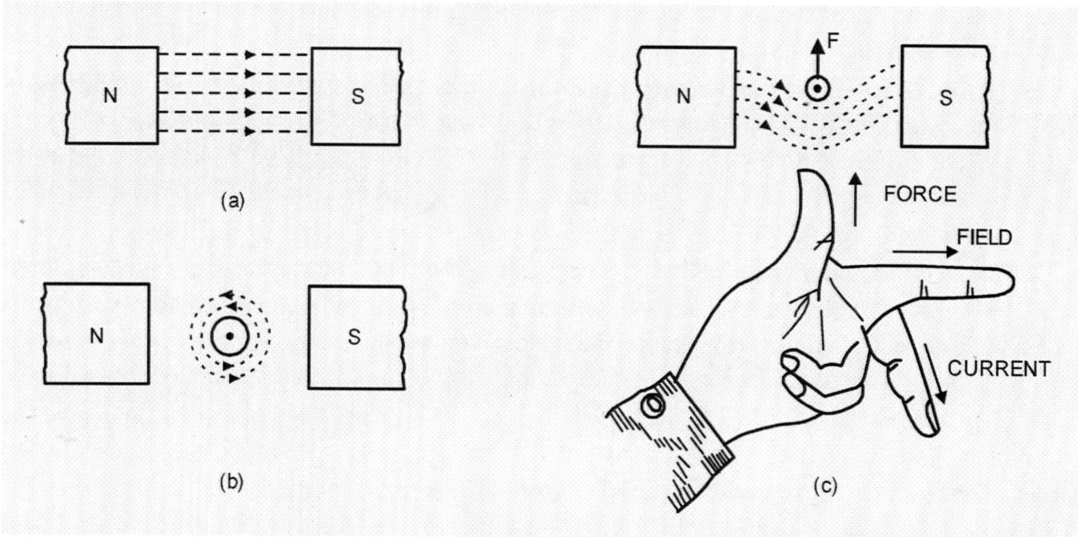

Fig. 6.14

Magnitude of force: When the conductor is placed at right angles to the uniform magnetic field, as shown in fig. 6.15 (*a*), the magnitude of mechanical force acting on the current carrying conductor is given by the relation;

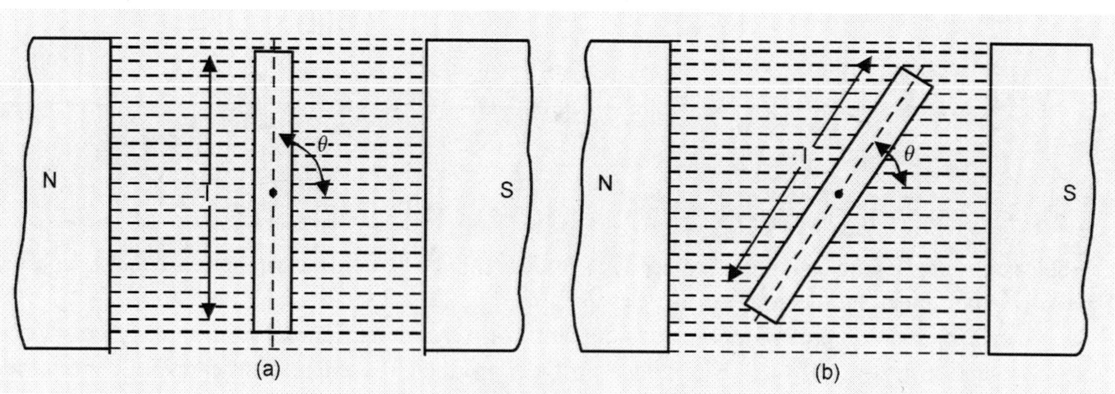

Fig. 6.15

$F = B\,I\,l$ newtons

where, B = flux density of the magnetic field in Wb/m^2;

I = current flowing through the conductor in ampere and

l = effective length of the conductor in metre.

However, when the conductor is not perpendicular to the magnetic field but is inclined making an angle $\theta\,(\theta < 90°)$ with the field as shown in fig. 6.15 (*b*). The magnitude of force acting on the conductor is reduced to;

$$F = B\,I\,l \sin\theta$$

This basic principle is applied for the production of torque (rotation) in electric motors, moving coil instruments etc.

Example 6.10. *A conductor, 20 cm long carrying a current of 25 A lies perpendicular to the field of strength 1000 AT/m. Calculate (i) force acting of the conductor; (ii) mechanical power required to move the conductor against the force with speed of 1 m/s; (iii) e.m.f. induced in the conductor.*

Solution: Force acting on the conductor, $F = B\,I\,l$

where, $B = \mu_0 H = 4\pi 10^{-7} \times 1000$ Wb/m^2; $I = 25$ A; $l = 0.2$ m;

$$\therefore \quad F = 4\pi\, 10^{-7} \times 1000 \times 25 \times 0.2$$
$$= \mathbf{6.283 \times 10^{-3}\ N}\ (Ans.)$$

Mechanical power required, $P = F \times$ velocity

$$= 6.283 \times 10^{-3} \times 1 = \mathbf{6.283 \times 10^{-3}\ W}\ (Ans.)$$

$$\text{Induced e.m.f., } e = \frac{P}{I} = \frac{62\cdot 83 \times 10^{-4}}{25} = \mathbf{0.2514 \times 10^{-3}\ V}\ (Ans.)$$

Example 6.11. *The armature winding of a d.c. motor consists of 250 turns (equivalent to 500 conductors). The effective length of each conductor in the field is 20 cm and the conductors are situated at a radius of 100 mm from the centre of the motor shaft. Each conductor carries a current of 15 A. Calculate the flux density which must be provided by the radial field in which the conductors lie, if the motor is to develop a torque to 180 Nm.*

Solution: We know, Torque = Force × radius

$$\therefore \quad \text{Total force, } F_t = 180/0.1 = 1800 \text{ N}$$

$$\text{Force produced by each conductor, } F = \frac{F_t}{\text{No. of conductors}} \qquad \therefore\ F = \frac{1800}{500} = 3.6 \text{ N}$$

Now, $F = B\,I\,l$ or $3.6 = B \times 15 \times 0.2$ or $B = \mathbf{1.2\ T}$ (*Ans.*)

6.16 MAIN CONSTRUCTIONAL FEATURES OF A DYNAMO

The complete assembly of various parts in a scattered form of a d.c. dynamo is shown in Fig. 6.16. The essential parts of a d.c. machine are described below:

1. Magnetic frame or Yoke

The outer cylindrical frame to which main poles and inter poles are fixed and by means of which the machine is fixed to the foundation is called the *yoke.* It serves two purposes:

(*i*) It provides mechanical protection to the inner parts of the machine.

(*ii*) It provides a low reluctance path for the magnetic flux.

The yoke is made of cast iron for smaller machines and for larger machines, it is made of cast steel or fabricated rolled steel since these materials have better magnetic properties as compared to cast iron.

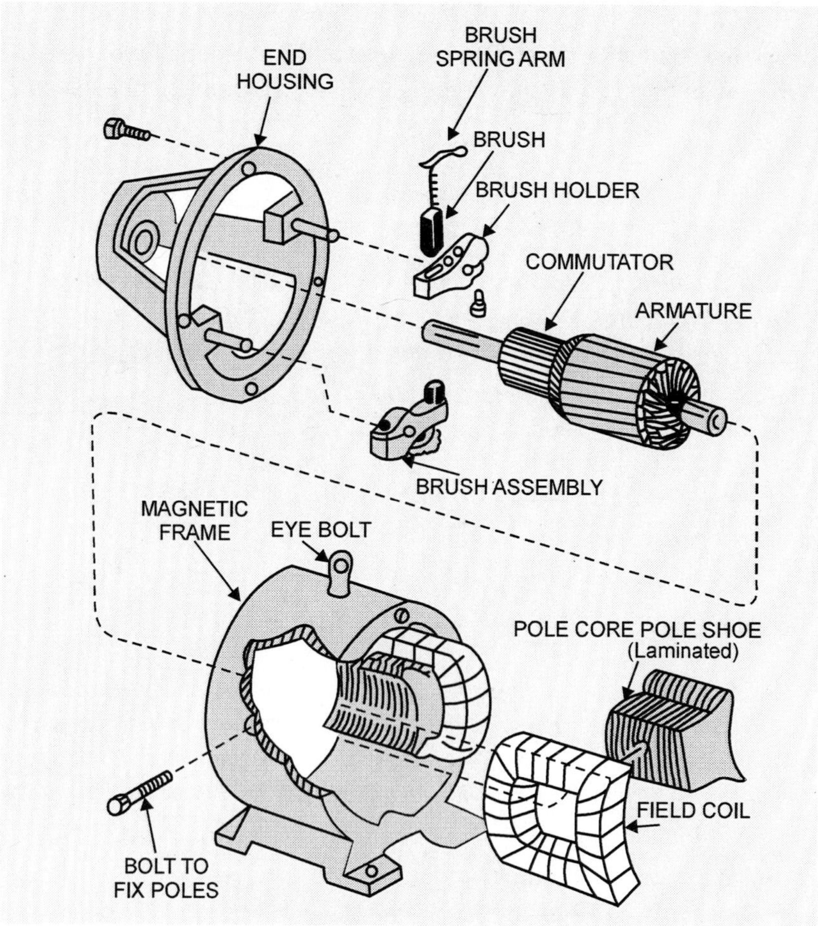

Fig. 6.16

2. Pole core and pole shoes

The pole core and pole shoes are fixed to the magnetic frame or yoke by bolts.

They serve the following purposes:

(*i*) They support the field or exciting coils.

(*ii*) They spread out the magnetic flux over the armature periphery more uniformly.

(*iii*) Since pole shoes have larger X-section, the reluctance of magnetic path is reduced.

Usually, the pole core and pole shoes are made of thin cast steel or wrought iron laminations which are riveted together under hydraulic pressure as shown in Fig. 6.17.

3. Field or Exciting coils

Enameled copper wire is used for the construction of field or exciting coils. The coils are wound on the former [see fig 6.18] and then placed around the pole core as shown in Fig. 6.17. When direct current is passed through the field winding, it magnetises the poles which produce the required flux. The field coils of all the poles are connected in series in such a way that when current flows through them, the adjacent poles attain opposite polarity as shown if Fig. 6.19.

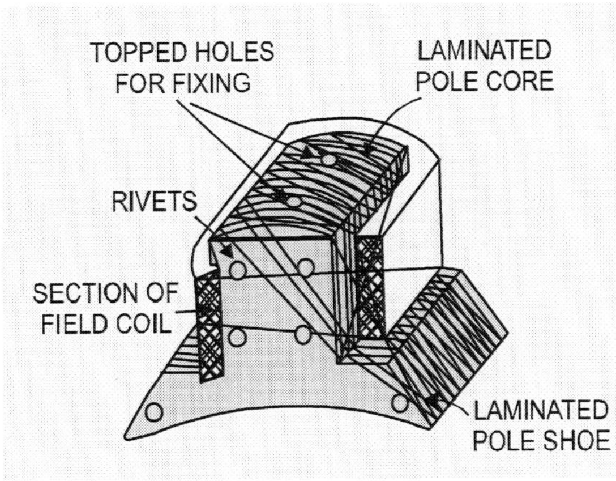

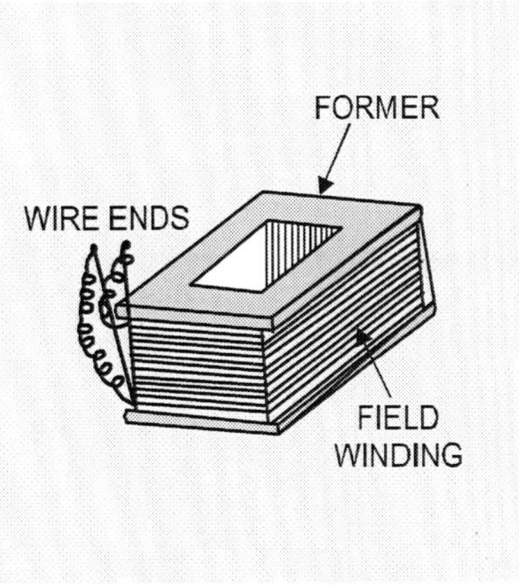

Fig. 6.17

Fig. 6.18

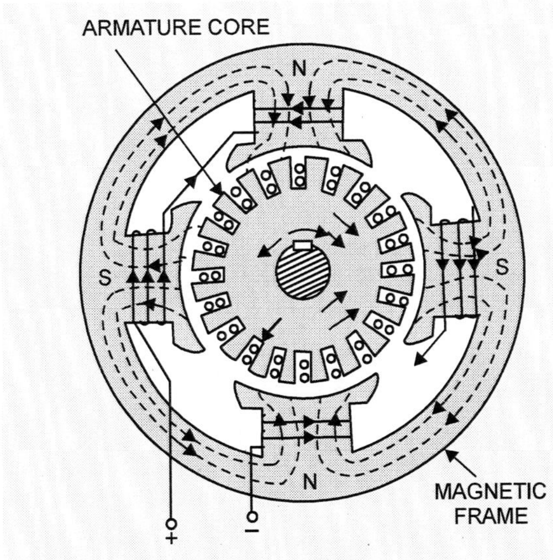

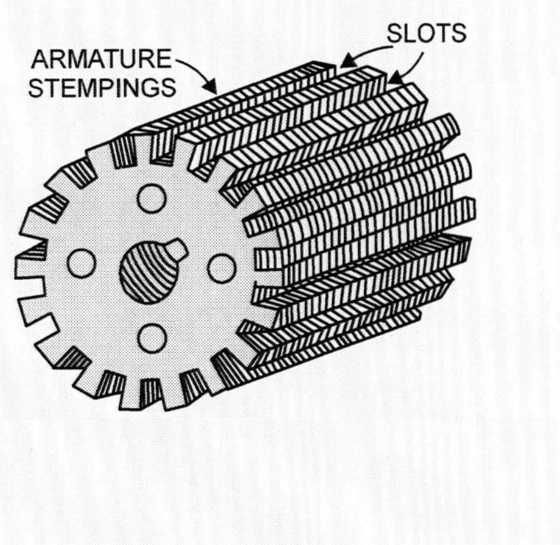

Fig. 6.19

Fig. 6.20

4. Armature core

It is cylindrical in shape and keyed to the rotating shaft. At the outer periphery slots are cut, as shown in Fig. 6.19, which accommodate the armature winding. The armature core shown in Fig. 6.20, serves the following purposes:

(*i*) It houses the conductors in the slots.

(*ii*) It provides an easy path for magnetic flux.

Since armature is a rotating part of the machine, reversal of flux takes place in the core, hence hysteresis losses are produced. To minimise these losses silicon steel material is used for its construction. The rotating armature cuts across the magnetic field which induces an e.m.f in it. This e.m.f. circulates eddy currents which results in eddy current loss in it. To reduce these losses, armature core is laminated, in other words we can say that about 0.3 to 0.5 mm thick stampings are used for its construction. Each lamination or stamping is insulated from the other by varnish layer (see Fig. 6.20).

5. Armature winding

The insulated conductors housed in the armature slots are suitably connected. This is known as armature winding. The armature winding is the heart of a dynamo. It is a place where conversion of power takes place *i.e.* in case of generator, mechanical power is converted into electrical power and in case of motor, electrical power is converted into mechanical power.

6. Commutator

It is the most important part of a d.c machine and serves the following purposes:

(*i*) It connects the rotating armature conductors to the stationary external circuit through brushes.

(*ii*) It converts the alternating current induced in the armature conductors into unidirectional current in the external load circuit in generator action, whereas, it converts the alternating torque into unidirectional (continuous) torque produced in the armature in motor action.

The commutator is of cylindrical shape and is made up of wedge-shaped hard drawn copper segments. The segments are insulated from each other by a thin sheet of mica. The segments are held together by means of two V-shaped rings that fit into the V-grooves cut into the segments. Each armature coil is connected to the commutator segment through riser. The sectional view of the commutator assembly is shown in Fig. 6.21.

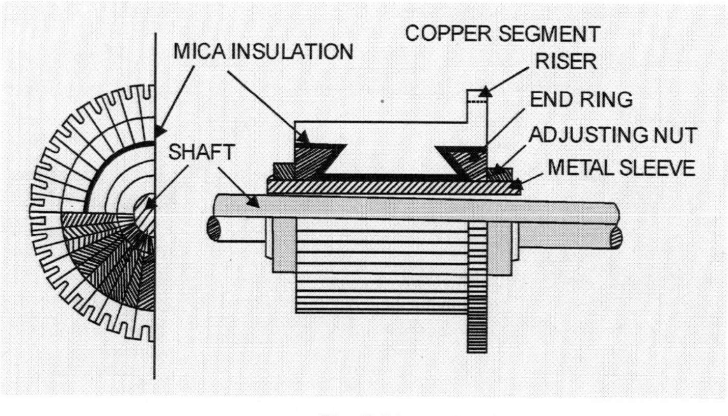

Fig. 6.21

7. Brushes

The brushes are pressed upon the commutator and form the connecting link between the armature winding and the external circuit. They are usually made of high grade carbon because carbon is conducting material and at the same time in powdered form provides imbricating effect on the commutator surface. The brushes are held in particular position around the commutator by brush holders and rocker.

8. End housings

End housings are attached to the ends of the main frame and support bearings. The front housing supports the bearing and the brush assemblies whereas the rear housing usually supports the bearing only.

9. Bearings

The ball or roller bearings are fitted in the end housings. The function of the bearings is to reduce friction between the rotating and stationary parts of the machine. Mostly high carbon steel is used for the construction of bearings as it is very hard material.

10. Shaft

The shaft is made of mild steel with a maximum breaking strength. The shaft is used to transfer mechanical power from or to the machine. The rotating parts like armature core, commutator, cooling fan etc. are keyed to the shaft.

6.17 SIMPLE LOOP GENERATOR AND FUNCTION OF COMMUTATOR

For simplicity, consider only one coil *AB* placed in the strong magnetic field. The two ends of the coil are joined to slip rings *A¢* and *B¢* respectively. Two brushes rest on these slip rings as shown in Fig. 6.22.

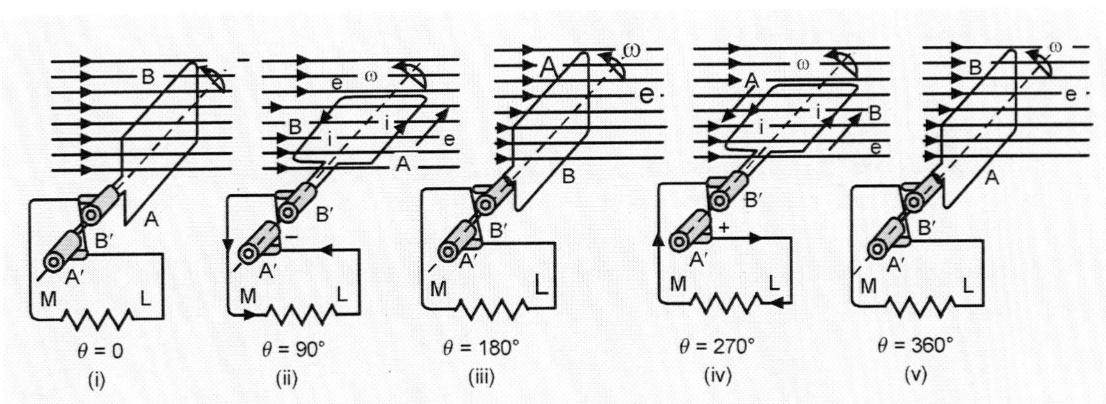

Fig. 6.22

When this coil is rotated in counter clockwise direction at an angular velocity of w radius per second, the magnetic flux is cut by the coil and an e.m.f. is induced in it. The position of the coil at various instants is shown in Fig. 6.22 and the corresponding value of the induced e.m.f. and its direction is shown in Fig. 6.23. The induced e.m.f. is alternating and the current flowing through the external resistance is also alternating *i.e.* at second instant current flows in external resistance from *M* to *L*, whereas, at fourth instant it flows from *L* to *M* as shown in Fig. 6.23.

Commutator action: Now, consider that the two ends of the coil are connected to only one slip ring splitted into two parts (segment) *i.e. A″* and *B″*. Each part is insulated from the other by a mica layer. Two brushes rest on these parts of the ring as shown in Fig. 6.24.

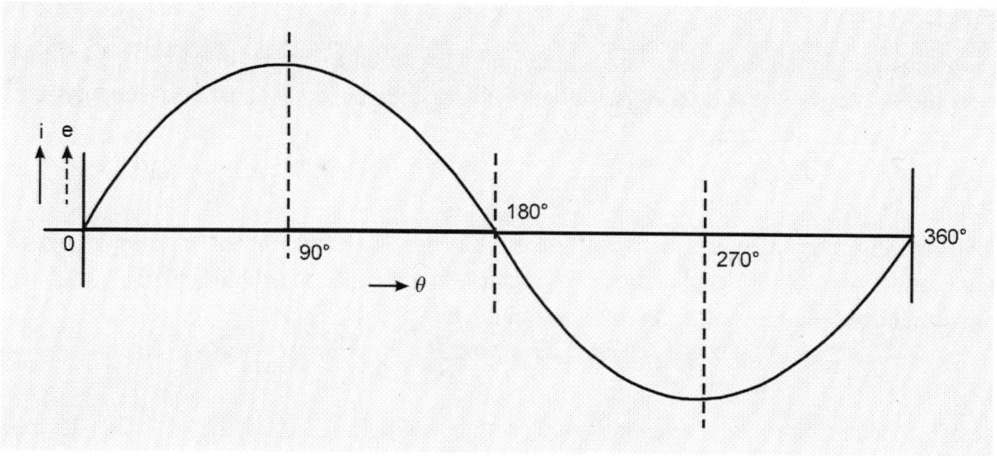

Fig. 6.23

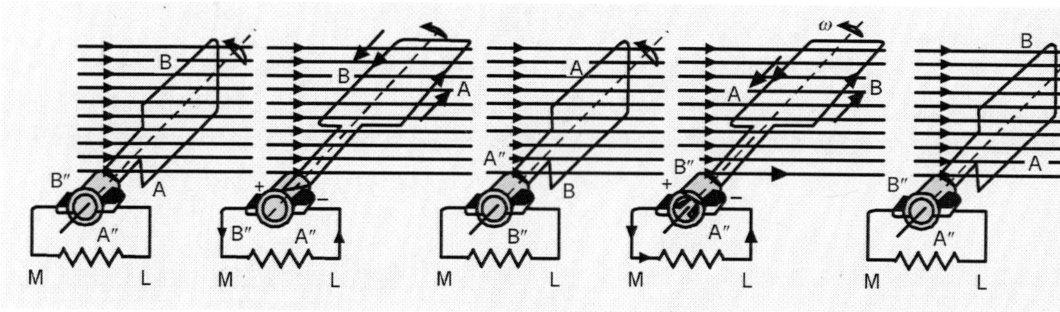

Fig. 6.24

In this case when the coil is rotated is counter clockwise direction at an angular velocity of *w* radians per second, the magnetic flux is cut by the coil and an e.m.f. is induced in it. The magnitude of e.m.f. induced in the coil at various instants will remain the same as shown in Fig. 6.25.

However, the flow of current in the external resistor or circuit will become unidirectional *i.e.* at second instant the flow of current in the external resistor is from *M* to *L* as well as the flow of current in the external resistor is from *M* to *L* in the fourth instant, as shown in Fig. 6.24. Its wave shape is shown in Fig. 16.25.

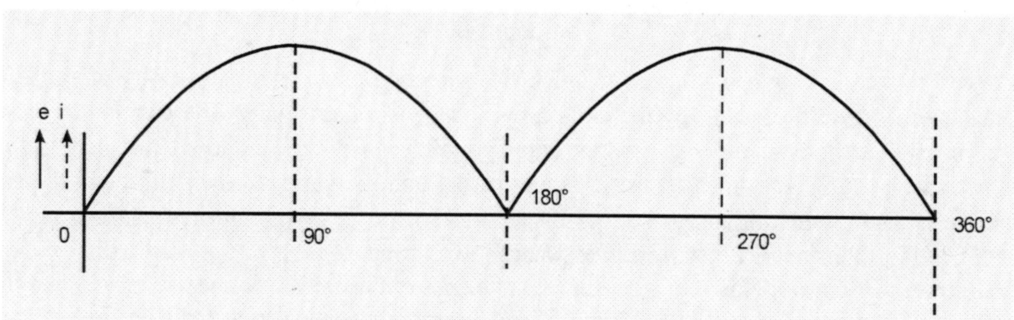

Fig. 6.25

Hence, *an alternating current is converted into unidirectional current in the external circuit with the help of a* **split ring** (*i.e.* **commutator**).

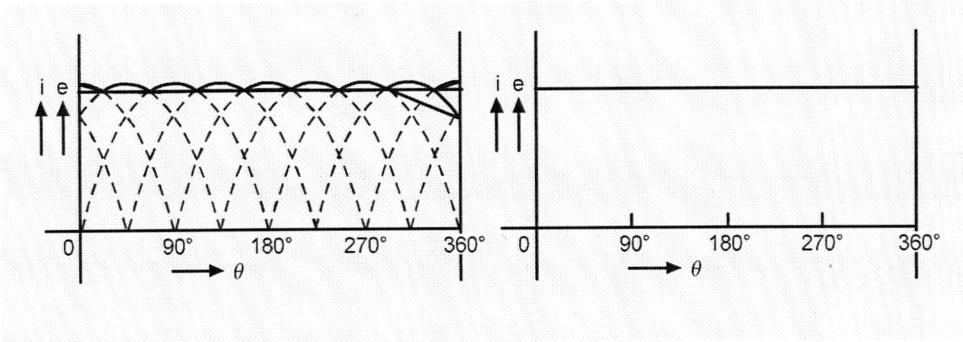

Fig. 6.26

In an actual machine, there are number of coils connected to the number of segments of the ring called commutator. The e.m.f. or current delivered by these coils to the external load is shown in Fig. 6.26 (*a*). The actual flow of current flowing in the external load is shown by the firm line which fluctuates slightly. The number of coils placed on the armature are even much more than this and a pure direct current is obtained at the output as shown in Fig. 6.26 (*b*).

Thus, *in actual machine working as a generator, the function of commutator is to convert the* **alternating current** *produced in the armature into* **direct current** *in the external circuit.*

TEST YOUR LEARNING-3

Type-I: Short Answer Questions

1. What will happen if a current carrying conductor is placed in a constant uniform magnetic field?
2. How will you determine the polarity of an electromagnet?
3. Why is the armature core of a dynamo is laminated?
4. In small dynamos, cast iron yokes are preferred, why?
5. What for field coils are provided in a dynamo?
6. What for brushes are employed in dynamos?
7. What is the function of armature in a dynamo?

Type-II: Numerical for Practice

1. A conductor, 30 cm long carrying a current of 20 A lies perpendicular to the field of flux density 1.885 m tesla. Calculate (i) force acting of the conductor; (ii) mechanical power required to move the conductor against the force with a speed of 0.75 m/s; (iii) e.m.f. induced in the conductor. (*Ans.* **11.31×10^{-3} N, 8.483×10^{-3} W, 0.424×10^{-3} V**)
2. The armature winding of a d.c. motor consists of 200 turns (equivalent to 400 conductors). The effective length of each conductor in the field is 25 cm and the conductors are situated

at a radius of 100 mm from the centre of the motor shaft. Each conductor carries a current of 20 A. Calculate the flux density which must be provided by the radial field in which the conductors lie, if the motor is to develop a torque to 200 Nm. (*Ans.* 1·0 T)

Type-III : Test Questions

1. Why force is exerted on a current carrying conductor placed in a uniform magnetic field?
2. State Fleming's Left Hande Rule.
3. Name major parts of a dynamo (dc). Draw the sketch and show the path of magnetic flux in a 4-pole dynamo.
4. What is the principle of operation of a dynamo? Why is a commutator and brush arrangement necessary for the operation of a dynamo as a dc generator?
5. With the help of sketches, describe the main parts and working principle of a dynamo.

SUMMARY

1. *Magnet:* A piece of iron that exibits magnetic field around it and has the tedancy to attract iron pieces is called a *magnet.*
2. *Pematent magnet:* If it retains these properties permanently then it is called a *pematent magnet.*
3. *Magnetic field* : The region around a magnet where its poles exhibit a force is called *magnetic field.*
4. Electromagnetism: *The phenomenon by which magnetism is produced by the effect of electric current is called* electromagnetism.
5. *Right hand thumb rule :* Hold the current carrying conductor in your right hand so that thumb points in the direction of flow of current. The fingers curling around the conductor will point in the direction of magnetic lines of force around it.
6. *Electromagnet:* A magnetic material when obtains the magnetic properties by providing a current carrying solenoid around it is called an *electromagnet.*
7. *Electro-magnetic induction* : The phenomenon by which an e.m.f. is induced in a coil or circuit when magnetic flux linking with it changes is called *electromagnetic induction.*
8. *Faraday's laws of electro-magnetic induction* :

   First law : When even magnetic flux linking with any circuit changes an e.m.f. in induced in it.

   Second law : The magnitude of induced e.m.f. is directly proportional to rate of change of flux linkages.
9. *Fleming's Right Hand Rule* : Stretch first finger, second finger and thumb of right hand mutually perpendicular to each other. If first finger represents magnetic field, thumb represents the direction of motion of conductor w.r.t. magnetic field then the second finger will represent the direction of induced e.m.f. in the conductor.
10. *Lenz's law* : In effect, electromagnetically induced e.m.f. always opposes the cause which produces it.
11. *Dynamically induced e.m.f.* : When either conductor or field system or both of them are moving

so that flux is cut by the conductor, the e.m.f. induced in the conductor by this way is called *dynamically induced e.m.f.* $e = B\ l\ u \sin q$.

12. *Self-inductance* : The property of a coil due to which it opposes the change of current flowing through itself is called *self inductance* of the coil.

$$L = \frac{e}{dI/dt};\ L = \frac{N\phi}{I};\ L = \frac{N^2}{l}\ a\ \mu_0\ \mu_r$$

13. *Mutual-inductance* : The property of one coil due to which it opposes the change of current flowing through the other (neighbouring) coil is called *mutual-inductance* between the two coils.

$$M = \frac{e_m}{dI_1/dt};\ M = \frac{N_2\,\phi_{12}}{I_1};\ M = \frac{N_1 N_2}{l} a\,\mu_0\,\mu_r$$

14. *Energy stored in the magnetic field* : W or $E = \frac{1}{2} L I^2$ joule

15. *Right Hand Thumb Rule*: It is used to determine the direction of magnetic field set-up around a current carrying conducror *i.e.*

Hold the current carrying conductor in your right hand in such a way that thumb is parallel to the conductor and indicates the direction of flow of current then the fingers curling around the conductor will represent the direction of magnetic field set-up around the conductor.

It is also used to determine the polarity of electro-magnet *i.e.*

Hold the solenoid wound around the core of the electro-magnet in your right hand in such a way that thumb is parallel to the core, if the fingers curling around point the direction of flow of current in the solenoid, the thumb will point in the direction of magnetic field set-up in the core *i.e.* North pole of the electro-magnet.

16. *Fleming's Right Hand Rule* : Stretch first finger, second finger and thumb of right hand mutually perpendicular to each other. If first finger represents magnetic field, thumb represents the direction of motion of conductor w.r.t. magnetic field then the second finger will represent the direction of induced e.m.f. in the conductor.

17. *Main parts of a dynamo*: The main parts of a dynamo are:

Magnetic frame or Yoke, Pole core and pole shoe, Field winding, Armature, Armature winding, Commutator, Brushes. End rings, Bearings, Shaft etc.

SELF ASSESMENT
TEST YOUR PREPARATION

OBJECTIVE TYPE QUESTIONS

(*i*) Fill in the blanks: *Fill-up the blanks in the following with suitable words/ numeral*

1. An iron piece that exibits magnetic field around it and has the tedancy to attract iron pieces is called a __________.

2. The phenomenon by which magnetism is produced by the effect of electric current is called __________.
3. The direction of magnetic lines of force around a current carrying conductor is determined by applying __________.
4. A magnetic material when obtains the magnetic properties by providing a current carrying solenoid around it is called an __________.
5. Faraday's first law of electro-magnetic induction states that "Whenever a conductor cuts across the magnetic field, __________ is induced in the conductor."
6. The direction of induced emf in a conductor when cuts across the magnetic field, is determined by __________ .
7. The mutual inductance is given by the relation $M = \dfrac{N_1 \ldots\ldots\ldots}{\ldots\ldots\ldots}$
8. The magnitude of electromagnetically induced e.m.f. is directly proportional to the rate of change of __________.
9. The property of a coil due to which it opposes the change of current in the neighbouring coil is called __________.
10. The magnitude of dynamically induced e.m.f. is given by the expression ; $e =$ __________
11. The energy stored in the magnetic field is __________ joule.
12. The direction of electromagnetically induced e.m.f. is determined by __________ or __________.
13. The property of a coil due to which it opposes the change of current through it is called __________.
14. The unit of self inductance is __________.
15. An inductance of 20 henry carrying a current of 600 A will store __________ kWh energy.

(ii) Multi-choice questions: *Select the correct answer in the following:*

1. *The direction of electro-magnetically induced e.m.f. is determined by*
   (a) Flemings' right hand rule (b) Lenz's law
   (c) Right hand thumb rule (d) both a and b.
2. *A coil of 0·02 mH is carrying a current of 1 A. If this current is reversed in 0·02 second, the induced e.m.f. in the coil will be*
   (a) 0·08 V (b) 0·004 V (c) 0·002 V (d) 0·02 V
3. *The mutual inductance is given by the relation*
   (a) $\dfrac{e_m}{d\,I_1/dt}$ (b) $\dfrac{N_2\,\phi_{12}}{I_1}$ (c) $\dfrac{N_1\,N_2}{l}\,a\,\mu_0\,\mu_r$ (d) all of these

NUMERICALS FOR PRACTICE

1. A current of 8 A through a coil of 300 turns produces a flux of 4 m Wb. If this current is reduced to 2 A in 0·1 second, calculate average e.m.f. induced in the coil, assuming flux to be proportional to current. (**Ans.** *9 V*)
2. A coil of 500 turns in wound on a magnetic circuit of reluctance 2×10^5 AT/Wb. If a current of 2 A flowing through the coil is reversed in 10 ms, find the average e.m.f. induced in the coil. (**Ans.** *300 V*)
3. Estimate the inductance of a solenoid of 2500 turns wound uniformly over a length of 0·5 m on a cylinderical paper tube 4 cm in diameter. The medium is air. (**Ans.** *19·74 m H*)
4. A toroid having 25 cm mean diameter and 6.25 cm^2 circular area of cross-section is wound uniformly with 1000 turns of wire. Determine (*i*) inductance of the toroid ; (*ii*) e.m.f. induced in the coil when current in it is increasing at the rate of 200 A/s and (*iii*) energy stored in its magnetic field when coil carries a current of 10 A. (**Ans.** *1 m H; 0·2 V ; 0·05 J*)
5. A current of 20 A is passed through a coil of self-inductance 800 m H. Find the magnetic energy stored. If the current is reduced to half, find the new value of energy stored and the energy released back to the electrical circuit. (**Ans.** *160 J, 40 J, 120 J*)

SHORT ANSWER TYPE QUESTIONS

1. What is a magnet ?
2. What do you understand by magnetism ?
3. What are permanent and temporary magnets ?
4. Give the advantages and practical applications of temporary and permanent magnets ?
5. What are magnetic poles ?
6. What are magnetic lines of force ?
7. What do you understand by magnetic field ?
8. Electromagnets are used as jaw of cranes, how ?
9. What do you understand by electromagnetic induction ?
10. Which material would you select for making electromagnets ?
11. Define Faraday's laws of electromagnetic induction.
12. State Lenz's law.
13. State Fleming Right hand rule as well as Fleming Left hand rule.
14. State and right hand rule which is used to find the direction of magnetic field and current.
15. State Right hand cork screw rule.
16. What happens when a current carrying conductor is placed in a magnetic field ?

17. What do you mean by dynamically induced emf ?
18. What do you understand by statically induced emf ?
19. Define self inductance and give its unit.
20. Define Mutual inductance and give its unit.
21. Distinguish between self induced and mutually induced emf.
22. What is the energy store in an inductor ?

TEST QUESTIONS

1. What do you understand by permanent and temporary magnets? What are the characteristics of magnetic lines of force? How will you make an electro-magnet?
2. State and explain Right Hand Thumb Rule.
3. State and explain Faraday's laws of electro-magnetic induction.
4. State and explain : (*i*) Fleming's right hand rule (*ii*) Lenz's law.
5. Write down the expression for dynamically induced e.m.f. developed in a conductor of length *l metre* moving with a velocity of *v metre per second* in a uniform magnetic flux density of B Wb/m^2.
6. What do you understand by self induced e.m.f. and mutually induced e.m.f. ?
7. Explain the term self and mutual inductance.
8. Distinguish between self and mutual inductance.
9. Differentiate between :

   (*i*) Statically and Dynamically induced emfs

ANSWERS TO OBJECTIVE TYPE QUESTIONS

(*i*) Fill in the blanks:

1. magnet	**2.** electromagnetism	**3.** right hand thumb rule
4. electromagnet	**5.** an emf	**6.** Flaming's right hand rule
7. $N_2/(l/a\ m_o m_r)$	**8.** fluxflinkages	**9.** mutual inductance
10. $B\,l\,u \sin q$	**11.** $\frac{1}{2} L I^2$	**12.** Flemings' right hand rule or Lenz's law
13. self inductance	**14.** henry	**15.** one

(*ii*) Multi-Choice questions:

1. (*d*) **2.** (*c*) **3.** (*d*)

CHAPTER

7 A.C. CIRCUITS

TOPICS TO BE DISCUSSED

Generation of A.C. voltage, its generation and wave shape. Cycle, frequency, peak value R.M.S. value, form factor, crest factor, Phase difference, power and power factor, A.C. Series Circuits with (i) resistance and inductance (ii) resistance and capacitance and (iii) resistance, inductance and capacitance, Q-factor of R.L.C. series circuits.

LEARNING OBJECTIVES

After the completion of this ***chapter,*** *the students/readers will be able to understand:*

1. What is alternating voltage and current and how it is different to dc?
2. How sinusoidal ac is produced?
3. Phasor representation of alternating quantities.
4. Terms used in ac system such as frequency, time period, cycle, alternation etc.
5. Instantaneous value, peak value, average value and rms value.
6. Behaviour of pure resistive, pure inductive and pure capacitive ac circuit.
7. Behaviour of *RL*, *RC* or *RLC* series circuits when connected across 1-phase ac supply.
8. Q-factor of *RLC* ac series circuits.

INTRODUCTION

In the previous chapters, we have dealt with the circuits in which the flow of current was steady and in one direction only *i.e.* direct current (*d.c.*). The use of d.c. is limited to a few applications *e.g.* electroplating, charging of batteries, electric traction, electronic circuits etc. However, for large scale power generation, transmission, distribution and utilisation a.c. system is invariably adopted. In a.c. system, voltage acting in the circuit changes polarity and magnitude at regular interval of time and hence the current. This system has the following important advantages over *d.c.* system:

(*i*) The alternating voltage can be stepped up and stepped down efficiently by means of transformer. To transmit huge power over a long distance, the voltages are stepped up (upto 400 kV) for economical reasons at the generating stations. Whereas, they are stepped down to a very low level (400/230 V) for utilisation of electrical energy from safety point of view.

Alternating supply is invariably used for domestic and industrial applications. The path for the flow of alternating current is called an **a.c. circuit**.In this chapter we shall deal with ac fundamentals and ac circuits.

7.1 ALTERNATING VOLTAGE AND CURRENT

A voltage that changes its polarity and magnitude at regular intervals of time is called an *alternating voltage*.

When an alternating voltage source is connected across a load resistor R, as shown in Fig. 7.1, the current flows through it in one direction and then in opposite direction when the polarity is reversed.

Fig. 7.1 (*c*) shows the wave shape of the source voltage (representing the variation of voltage w.r.t. time) and current flowing through the circuit (*i.e.* load resistor *R*).

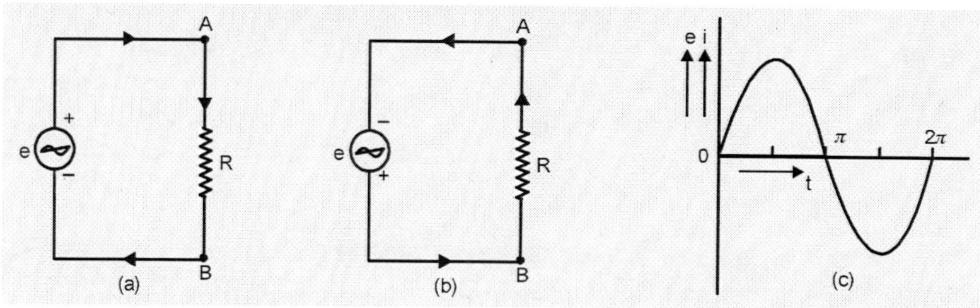

Fig. 7.1

Wave form: The graph representing the manner in which an alternating voltage or current changes w.r.t. time is known as *wave-form* or *wave-shape*. While plotting a graph, usually the instantaneous values of the alternating quantities are taken along *y-axis* and time along *x-axis*. The alternating voltage or current may vary in different manner, as shown in Fig. 7.2.

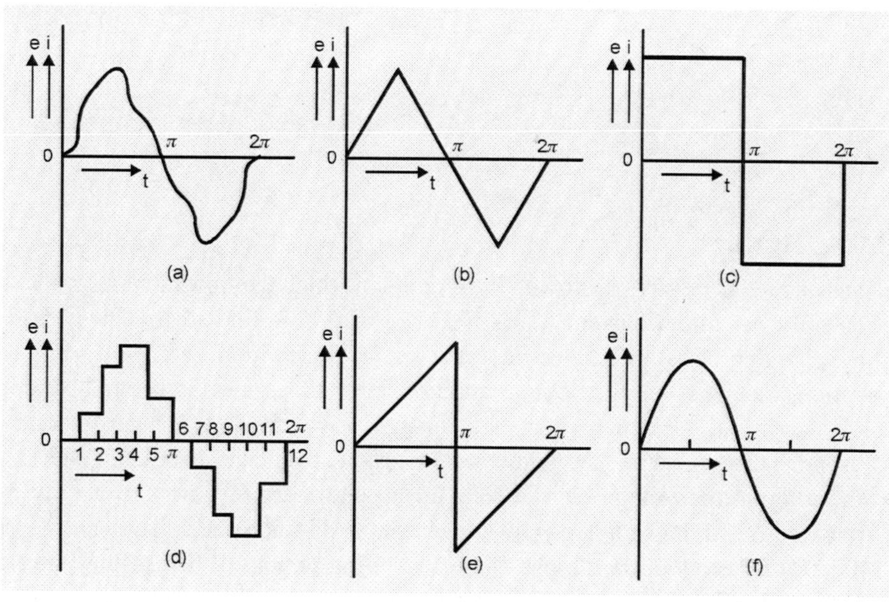

Fig. 7.2

7.2 DIFFERENCE BETWEEN AC AND DC

AC current	*DC* current
1. Alternating current reverses periodically and its magnitude changes.	1. Direct current flows only in one direction and remains unaltered.
2. Amplitude and polarities are varying continuously.	2. Amplitude and polarities are fixed.
3. It has a particular frequency.	3. It is independent of frequency.
4. *AC* can be generated at higher voltages.	4. *DC* cannot be generated at high voltages because of commutation difficulties.
5. In case of *ac*, the cost of generation is less.	5. In case of *dc*, the cost of generation is more.
6. Alternating voltage can be increased (stepped up) or decreased (stepped down) easily with the help of a transformer.	6. Direct voltage cannot be increased or decreased easily.

7.3 GENERATION OF ALTERNATING VOLTAGE AND CURRENT

An alternating voltage can be generated either

(*i*) by rotating a coil in a uniform magnetic field at constant speed as shown in Fig. 7.3. or

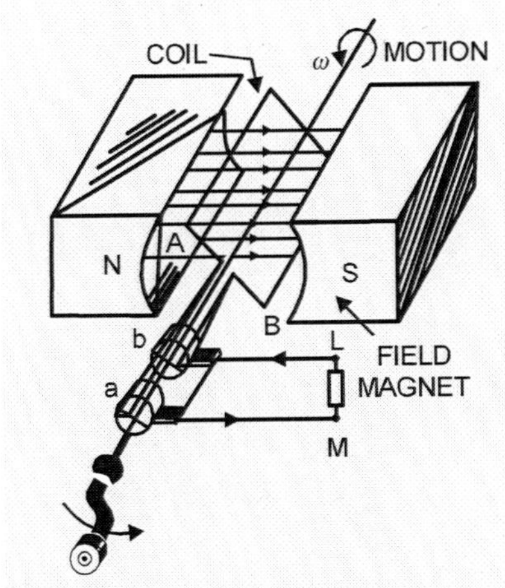

Fig. 7.3

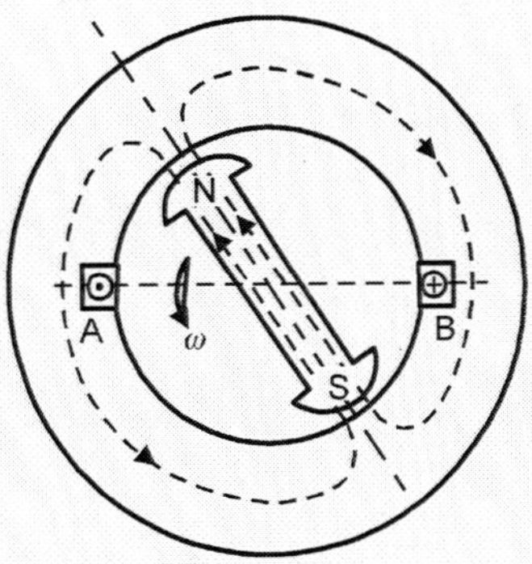

Fig. 7.4

(*ii*) by rotating a uniform magnetic field within a stationary coil at a constant speed as shown in Fig. 7.4.

The first method is generally applied in small a.c. generators, whereas, second method is applied in large a.c. generators due to economical considerations. In both the cases, magnetic field is cut by the conductors (or coil sides) and an e.m.f. is induced in them. The direction and magnitude of the induced e.m.f. in the conductors depend upon the position of the conductors as explained below:

For simplicity, consider a coil placed in a uniform magnetic field to which a load (*LM*) is connected through brushes and slip rings as shown in Fig. 7.5. When it is rotated in anticlockwise direction at a constant angular velocity of ω radians per second, an e.m.f. is induced in the coil sides. The cross-sectional view of the coil and its different positions at different instants are shown in Fig. 7.5.

The magnitude of induced e.m.f. depends upon the rate at which the flux is cut by the conductors. At (*i*), (*iii*) and (*v*) instants, induced e.m.f. in the conductor A and B is zero as they are moving parallel to the magnetic lines of force and the rate of flux cut is zero. Whereas, the magnitude of e.m.f. induced in the conductor A and B is maximum at instant (*ii*) and (*iv*) as the conductors are moving perpendicular to the magnetic lines of force and the rate of flux cut is maximum.

The direction of e.m.f. induced in the conductors is determined by applying *Fleming's right hand rule*. At instant (*ii*), the direction of e.m.f. induced in conductor A is outward whereas, at instant (*iv*), the direction of induced e.m.f. in the conductor A is inward (*i.e.* the direction of induced e.m.f. at this instant is opposite to that of the direction of induced e.m.f. at instant (*ii*).

The wave shape of the e.m.f. induced in the coil is also shown in Fig. 7.5.

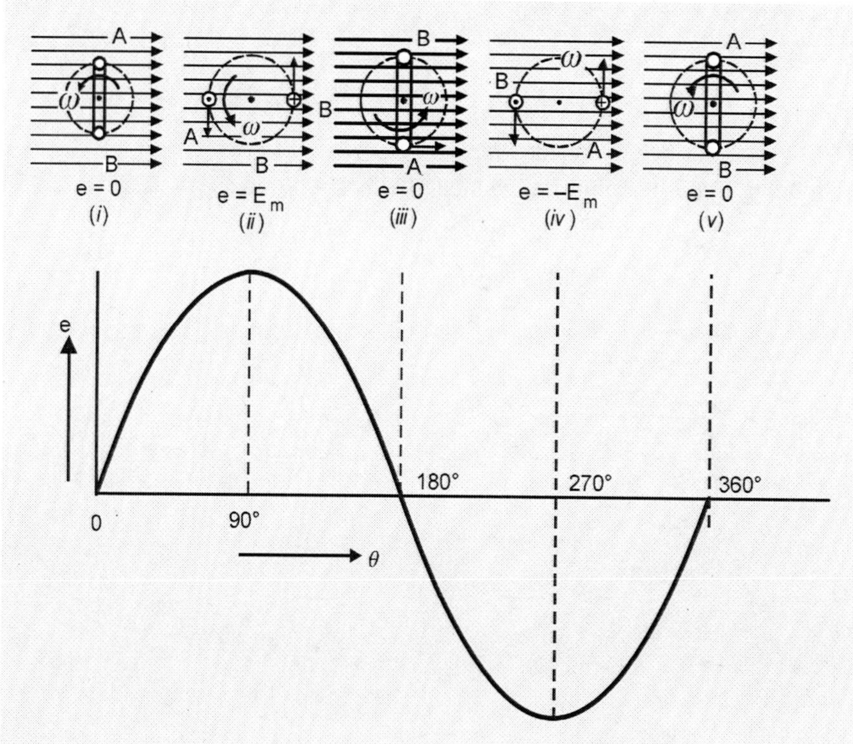

Fig. 7.5

7.4 EQUATION OF ALTERNATING E.M.F. AND CURRENT

Consider a coil having N turns rotating in a uniform magnetic field of density B Wb/m^2 in the counter-clockwise direction at an angular velocity of ω radians per second as shown in Fig. 7.6.

At the instant, as shown in Fig. 7.6 (*b*) maximum flux ϕ_m is linking with the coil. After t seconds, the coil is rotated through an angle $\theta = \omega t$ radians. The component of flux linking with he coil at this instant is ϕ_m cos ωt. Whereas, the other component ϕ_m sinωt is parallel to the plane of the coil.

According to Faraday's laws of electromagnetic induction, the magnitude of e.m.f. induced in the coil at this instant *i.e.*

Instantaneous value of e.m.f. induced in the coil,

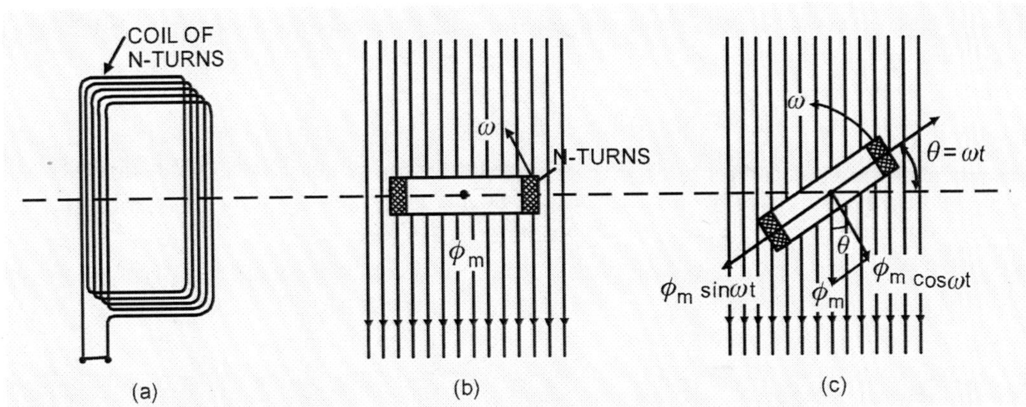

Fig. 7.6

$$e = -N\frac{d\phi}{dt}$$ (–ve sign indicates that in effect the induced e.m.f. is opposite to the very cause which produces it)

or $$e = -N\frac{d}{dt}\phi_m \cos\omega t \quad (\because \phi = \phi_m \cos \omega t)$$

or $$e = -N\,\phi_m(-\omega \sin \omega t)$$

or $$e = \omega N \phi_m \sin \omega t \qquad ...(i)$$

The value of induced e.m.f. will be maximum when angle θ or $\omega t = 90°$ (*i.e.* $\sin \omega t = 1$)

$$\therefore \quad E_m = \omega N \phi_m \qquad ...(ii)$$

Putting this value in equation (*i*), we get,

$$\mathbf{e = E_m \sin \omega t = E_m \sin \theta}$$

From the above equation it is clear that the magnitude of the induced e.m.f. varies according to sine of angle θ. The wave shape of the induced e.m.f. is shown in Fig. 7.7. This wave form is called sinusoidal wave.

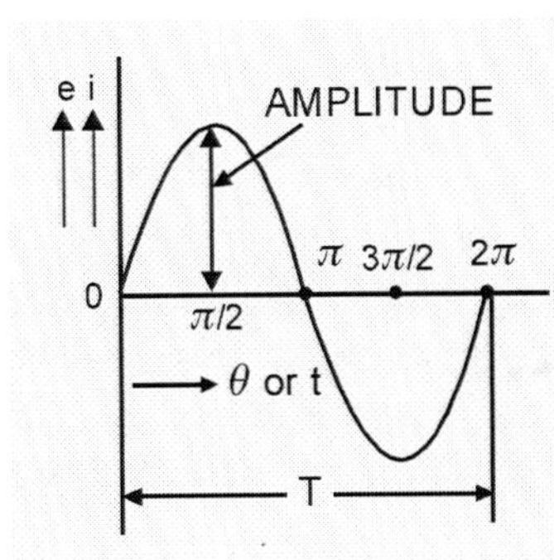

Fig. 7.7

If this voltage is applied across resistor, an alternating current will flow through it varying sinusoidally *i.e.* following a sine law and its wave shape will be same as shown in Fig. 7.7.

This alternating current is given by the equation:

$$i = I_m \sin \omega t = I_m \sin \theta$$

7.5 IMPORTANT TERMS

An alternating voltage or current changes its magnitude and direction at regular intervals of time. A sinusoidal voltage or current varies as a sine function of time *t* or angle *q* (= *w t*). The following important terms are generally used in alternating quantities:

(*i*) **Wave form:** The shape of the curve obtained by plotting the instantaneous values of alternating quantity (voltage or current) along *y*-axis and time or angle ($q = w\ t$) along *x*-axis is called *wave form* or *wave shape*. Fig. 7.7 shows the waveform of an alternating quantity varying sinusoidally.

(*ii*) **Instantaneous value:** The value of an alternating quantity *i.e.* voltage or current at any instant is called its *instantaneous value* and is represented by *e* or *i* respectively.

(*iii*) **Cycle:** When an alternating quantity goes through a complete set of +ve and –ve values or goes through 360 electrical degrees, it is said to have completed one *cycle*.

(*iv*) **Alternation:** One half cycle is called *alternation*. An alternation spans 180 electrical degrees.

(*v*) **Time period:** The time taken in seconds to complete one cycle by an alternating quantity is called *time period*. It is generally denoted by *T*.

(*vi*) **Frequency:** The number of cycles made per second by an alternating quantity is called *frequency*. It is measured in cycles per second (c/s) or hertz (Hz) and is denoted by *f*.

(*vii*) **Amplitude:** The maximum value (positive or negative) attained by an alternating quantity in one cycle is called its *amplitude* or *peak value* or *maximum value*. The maximum value of voltage and current is generally denoted by E_m (or V_m) and I_m respectively.

7.6 IMPORTANT RELATIONS

Some of the terms used in a.c. terminology have definite relations among themselves as given below:

(*i*) **Relation between frequency and time period:** Consider an alternating quantity having a frequency of *f* c/s. Then

Time taken to complete *f* cycle = 1 s

Time taken to complete 1 cycle = $1/f$ second

Hence time period, $T = 1/f$ second   or $f = 1/T$ c/s

(*ii*) **Relation between frequency and angular velocity:** Consider an alternating quantity having a frequency of *f* c/s.

Angular distance covered in one cycle $= 2\pi$ radian

∴ Angular distance covered per second in *f* cycles $= 2\pi$ radian

Hence, $\omega = 2\ \pi f$ radian/s

7.7 VALUES OF ALTERNATING VOLTAGE AND CURRENT

The voltage and current in d.c. system are constant so that there is no problem of specifying their magnitudes. Whereas, in a.c. system, the alternating voltage and current varry from instant to instant. So the question arises how to express the magnitude of alternating voltage and current. The following three ways are adopted to express the magnitude of these quantities:

(*i*) Peak Value.

(*ii*) Average value or Mean value.

(*iii*) Effective value or R.M.S. value.

The r.m.s. value of an alternating quantity (voltage or current) represents the real magnitude. Whereas, the peak and average values are important in some of the engineering applications.

7.8 PEAK VALUE

The maximum value attained by an alternating quantity during one cycle is called is *peak value*. This is also called *maximum value* or *crest value* or *amplitude*. The peak of an alternating voltage and current is represented by E_m and I_m.

7.9 AVERAGE VALUE

The arithmetic average of all the instantaneous values considered of an alternating quantity (current or voltage) over one cycle is called **average value.**

In case of symmetrical waves (like sinusoidal current or voltage wave) the +ve half is exactly equal to the –ve half, therefore, the average value over a complete cycle is zero. Since work is being done by the current in the +ve as well as in –ve half cycle, therefore, average value is determined regardless of signs. Hence, to determine average value of alternating quantities having symmetrical waves, only (+ve half) cycle is considered.

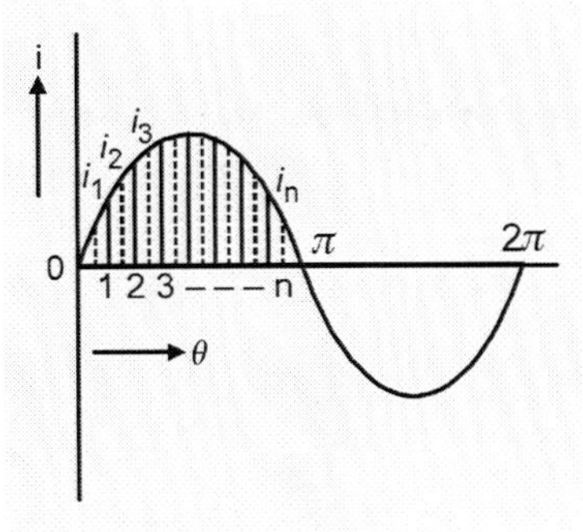

Fig. 7.8

Divide the +ve half cycle into n number of equal parts as shown in Fig. 7.8. Let $i_1, i_2, i_3 .... i_n$ be the mid-ordinates.

Average value of current, I_{av} = mean of mid-ordinates.

$$= \frac{i_1 + i_2 + i_3 + ... + i_n}{n}$$

$$= \frac{\text{Area of alternation}}{\text{base}}$$

7.10 AVERAGE VALUE OF SINUSOIDAL CURRENT

The alternating current varying sinusoidally, as shown in Fig. 7.9, is given by the equation:

$$i = I_m \sin\theta$$

Consider an elementary strip of thickness $d\theta$ in the +ve half cycle, i be its mid-ordinate. Then

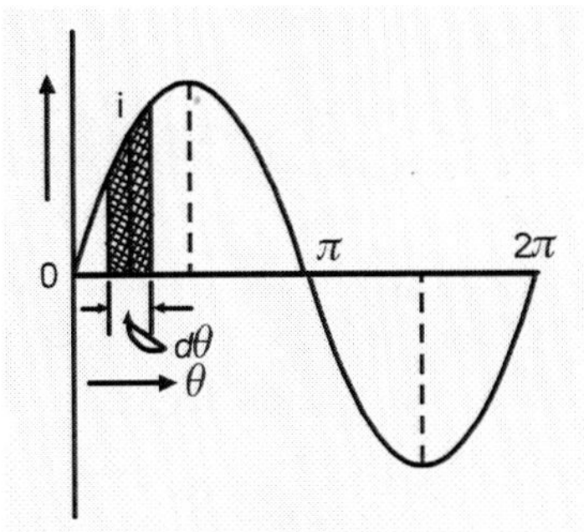

Fig. 7.9

$$\text{Area of strip} = i\, d\theta$$

$$\text{Area of half cycle} = \int_0^{\pi} i\, d\theta = \int_0^{\pi} I_m \sin\theta\, d\theta$$

$$= I_m \left(-\cos\theta\right)_0^{\pi} = I_m \left(-(\cos\pi - \cos 0)\right)$$

$$= I_m\,[-(-1-1)] = 2I_m$$

Base = 0 to $\pi = \pi - 0 = \pi$

$$\therefore \text{Average value, } I_{av} = \frac{\text{Area of alternation}}{\text{base}} = \frac{2I_m}{\pi} = \frac{I_m}{\pi/2} = 0\cdot637\ I_m$$

7.11. EFFECTIVE OR R.M.S. VALUE

That steady current which when flows through a resistor of known resistance for a given time produces the same amount of heat as produced by the alternating current when flows through the same resistor for the same time is called **effective or r.m.s. value** *of the alternating current.*

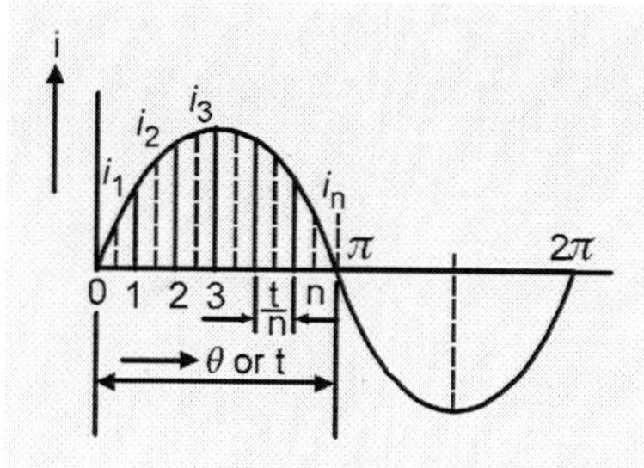

Fig. 7.10

Let i be the alternating current flowing through a resistor of resistance R for time t seconds which produces the same amount of heat as produced by I_{eff} (a direct current). The base of one alternation is divided into n equal parts, as shown in Fig. 7.10, so that interval is of $\frac{t}{n}$ second. Let $i_1, i_2, i_3, \ldots\ldots i_n$ be the mid-ordinate.

Then heat produced in the

First interval $= i_1^2 Rt / Jn$ calorie

Second interval $= i_2^2 Rt / Jn$ calorie

Third interval $= i_3^2 Rt / Jn$ calorie

nth interval $= i_n^2 Rt / Jn$ calorie

$$\text{Total heat produced} = \frac{Rt}{J}\left(\frac{i_1^2 + i_2^2 + i_3^2 + \ldots\ldots + i_n^2}{n}\right) \text{ calorie} \qquad \ldots(i)$$

Since I_{eff} is considered as the effective value of this current.

$$\text{Then total heat produced by this current} = I_{eff}^2\, Rt / J \text{ calorie} \qquad \ldots(ii)$$

Equating equation (*i*) and (*ii*), we get,

$$\frac{I_{eff}^2\, Rt}{J} = \frac{Rt}{J}\left(\frac{i_1^2 + i_2^2 + i_3^2 + \ldots .\, i_n^2}{n}\right)$$

or $$I_{eff} = \sqrt{\frac{i_1^2 + i_2^2 + i_3^2 + \ldots . + i_n^2}{n}} = \sqrt{\text{mean of squares of instantaneous values}}$$

or I_{eff} = Square root of mean of squares of instantaneous values

= root-mean-square (*r.m.s.*) value

It is the actual value of an alternating quantity which tells us the energy transfer capability of an a.c. source. For example, if we say that 5A alternating current is flowing through a circuit, it means the *r.m.s.* value of alternating current which flows through the circuit is 5A. It transfers the same amount of energy as is transferred by 5A d.c.

The ammeters and voltmeters record the *r.m.s.* values of alternating currents and voltages respectively. The domestic single-phase a.c. supply is 230 V, 50 Hz. Where 230 V is the *r.m.s.* value of alternating voltage.

7.12 R.M.S. VALUE OF SINUSOIDAL CURRENT

An alternating current varying sinusoidally is given by the equation;

$$i = I_m \sin\theta$$

To determine the r.m.s. value, the squared wave of the alternating current is drawn as shown in Fig. 7.11.

Considering an elementary strip of thickness $d\theta$ in the first half cycle of the squared wave. Let i^2 be its mid-ordinate. Then

$$\text{Area of strip} = i^2\, d\theta$$

Area of first half cycle of squared wave

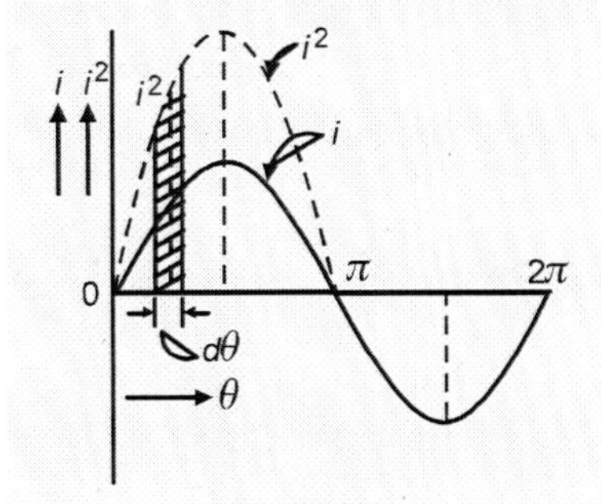

Fig. 7.11

$$= \int_0^{\pi} i^2\, d\theta = \int_0^{\pi} (I_m \sin\theta)^2\, d\theta$$

$$= I_m^2 \int_0^{\pi} \sin^2\theta\, d\theta = I_m^2 \int_0^{\pi} \frac{1-\cos 2\theta}{2} d\theta$$

$$= \frac{I_m^2}{2} \int_0^{\pi} (1-\cos 2\theta)\, d\theta = \frac{I_m^2}{2}\left(\theta - \frac{\sin 2\theta}{2}\right)_0^{\pi}$$

$$= \frac{I_m^2}{2}\left((\pi - 0) - \frac{\sin 2\pi - \sin 0}{2}\right) = \frac{I_m^2}{2}[(\pi - 0) - (0-0) = \frac{\pi I_m^2}{2}$$

Base $= 0$ to $\pi = \pi - 0 = \pi$

Effective or r.m.s. value,

$$I_{r.m.s.} = \sqrt{\frac{\text{Area of first half of squared wave}}{\text{base}}} = \sqrt{\frac{\pi I_m^2}{2\pi}} = \sqrt{\frac{I_m^2}{2}} = \frac{I_m}{\sqrt{2}} = 0.707\, I_m$$

Usually r.m.s. value of alternating current is simply represented by I instead of $I_{r.m.s.}$ Similarly, *r.m.s.* value of alternating voltage is represented by E or V.

7.13 FORM FACTOR AND PEAK FACTOR

There exists a definite relation among the average value, r.m.s. value and peak value of an alternating quantity. The relationship is expressed by the two factors namely, form factor and peak factor.

(*i*) Form factor: The ratio of r.m.s. value to average value of an alternating quantity is called *form factor*.

Mathematically, Form factor $= \frac{I_{r.m.s.}}{I_{av}}$ or $\frac{E_{r.m.s.}}{E_{av}}$

For the current varying sinusoidally;

$$\text{Form factor} = \frac{I_{r.m.s.}}{I_{av}} = \frac{I_m/\sqrt{2}}{2 I_m/\pi} = \frac{\pi I_m}{2\sqrt{2}\, I_m} = 1.11$$

(*ii*) Peak factor: The ratio of maximum value to r.m.s. value of an alternating quantity is called *peak factor*.

Mathematically, Peak factor $= \frac{I_m}{I_{r.ms.}}$ or $\frac{E_m}{E_{r.m.s.}}$

For current varying sinusoidally:

$$\text{Peak factor} = \frac{I_m}{I_{r.m.s.}} = \frac{I_m}{I_m/\sqrt{2}} = \sqrt{2} = 1.4142$$

Example 7.1. *The equation of an alternating current is* $i = 42.42 \sin 628\, t$

Determine: (*i*) its maximum value; (*ii*) Frequency; (*iii*) rms value; (*iv*) Average value; and (*v*) Form factor.

Solution: Given equation is

$$i = 42.42 \sin 628\, t$$

Comparing above equation with the standard equation, $i = I_m \sin \omega t$

$$i_m = 42.42 \text{ A and } \omega = 628 \text{ rad/sec}$$

(*i*) Maximum value, $I_m = 42.42$ **A** (*Ans.*)

(*ii*) Here, $\omega = 628$ rad/sec. or $2\pi f = 628$

$$\text{Frequency, } f = \frac{628}{2\pi} = \mathbf{100\ Hz}\ (Ans.)$$

(*iii*) $$\text{RMS value, } I_{rms} = \frac{I_m}{\sqrt{2}} = \frac{42 \cdot 42}{\sqrt{2}} = \mathbf{30\ A}\ (Ans.)$$

(*iv*) $$\text{Average value, } I_{av} = \frac{2 I_m}{\pi} = \frac{2 \times 42 \cdot 42}{\pi} = \mathbf{27\ A}\ (Ans.)$$

(*v*) $$\text{From factor} = \frac{I_{rms}}{I_{av}} = \frac{30}{27} = \mathbf{1.11}\ (Ans.)$$

Example 7.2. *A supply voltage of 230 V, 50 Hz is fed to a residential building. Write down its equation for instantaneous value.*

Solution: $v = V_m \sin \omega t$

where, $V_m = \sqrt{2}\, V_{r.m.s.} = \sqrt{2} \times 230 = 325.27$ V

$\omega = 2\pi f = 2\pi \times 50 = 314.16$ radians

$\therefore$ $v = \mathbf{325.27 \sin 314.16}\, t$ (*Ans.*)

Example 7.3. *An alternating voltage is given by* $v = 141.1 \sin 314t$. Find

(*i*) *Frequency*

(*ii*) *r.m.s. value*

(*iii*) *average value*

(*iv*) *the instantaneous value of voltage when 't'* is 3 *m sec.*

(*v*) *the time taken for the voltage to reach* 100V *for the first time after passing through zero value.*

Solution: (*i*) Given that, $v = 141.14 \sin 314\, t$

Comparing above equation with

$$v = V_{in} \sin \omega t \quad and \quad \omega = 314 \text{ rad/sec.}$$

$$\therefore \quad \text{Frequency, } f = \frac{314}{2\pi} = \frac{314}{2 \times 3 \cdot 14} = \mathbf{50\ Hz}\ (Ans.)$$

(*ii*) RMS value of the voltage, $$V_{rms} = \frac{V_m}{\sqrt{2}} = \frac{141 \cdot 4}{1 \cdot 414} = \mathbf{100\ V}\ (Ans.)$$

(*iii*) $$\text{Average value, } V_{av} = \frac{2 V_m}{\pi} = \frac{2 \times 141 \cdot 4}{\pi} = \mathbf{90\ V}\ (Ans.)$$

(*iv*) $v = 141.4 \sin 314\, t$. At $t = 3 \text{ m s} = 3 \times 10^{-3}$ s

$$v = 141.4 \sin 314 \times 3 \times 10^{-3} = \mathbf{114.4\ V}\ (Ans.)$$

(v) $v = 141.4 \sin 314\, t$ or $100 = 141.4 \sin 314\, t$

or $$t = \frac{1}{314} \sin^{-1}\left(\frac{100}{141 \cdot 4}\right) = 2.5 \times 10^{-3}\ \text{s} = \mathbf{2.5\ m\ s}\ (\textit{Ans.})$$

TEST YOUR LEARNING-1

Type-I: Short Answer Questions

1. Define an alternating quantity.
2. What do you mean by sinusoidal alternating quantity?
3. How the instantaneous value of a sinusoidal quantity is represented mathematically?
4. Define wave form of an alternating quantity.
5. Define a cycle of an alternating quantity.
6. Define alternation, time period, and frequency of an alternating quantity.
7. Give the relation between frequency and time period; frequency and angular velocity.
8. Define instantaneous and peak value of an alternating quantity.
9. What do you mean by an average value of an alternating quantity?
10. How will you define rms value of an alternating quantity.
11. What is form factor and peak factor.

Type-II: Numerical for Practice

1. A current is given by $i = 22.62 \sin 377\, t$, find (*i*) maximum value of current; (*ii*) r.m.s. value of current; (*iii*) frequency of current; (*iv*) radians through which its phasor has advanced after 0.01 second; (*v*) number of degrees in (*iv*) and (*vi*) value of current at instant mentioned in part (*iv*). (**Ans.** *22.62 A; 16 A; 60 Hz; 3.77 rad.; 216°; –13.3 A*)
2. A sinusoidal voltage of 50 Hz has a maximum value of $200\sqrt{2}$ V. At what time measured from a positive maximum value will the instantaneous voltage be 141.4 V? (**Ans.** *3.33 ms*)

Type-II: Test Questions

1. Derive a relation between average value and maximum value for a sinusoidal quantity.
2. Derive a relation between rms value and maximum value for a sinusoidal quantity.
3. Derive a relation between rms and average value of a sinusoidal quantity.
4. Derive a relation for peak factor of a sinusoidal quantity.

7.14 PHASOR REPRESENTATION OF SINUSOIDAL QUANTITY

It has been seen that an alternating quantity (varying sinusoidally) can be represented in the form of wave and equation. The wave form shows the graphical representation, whereas, the equation represents the mathematical expression of the instantaneous value of an alternating quantity.

The same alternating quantity can also be represented by a line of definite length (representing its maximum value) rotating in counter-clockwise direction at a constant velocity (ω radians/second). Such a rotating line is called a *phasor*.

Thus, an alternating quantity can be represented by a phasor which shows its magnitude and direction at that instant, as shown in Fig. 7.12.

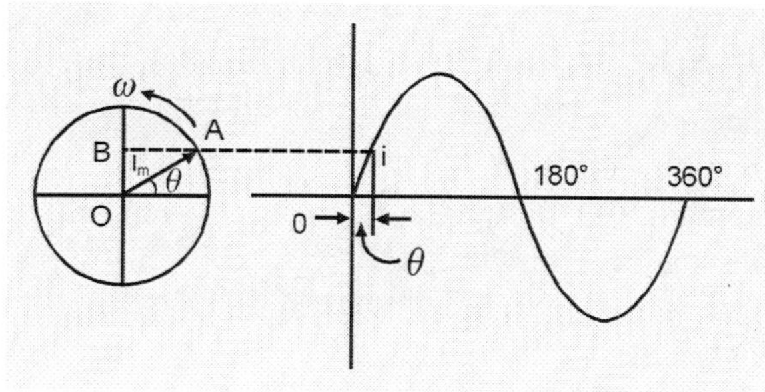

Fig. 7.12

The phasor representation of an alternating quantity enables us to understand its magnitude and position on the axis. The alternating quantities can be added and subtracted with a fair degree of ease by representing them vectorially (phasor diagram).

7.15 PHASE DIFFERENCE

The two alternating quantities having same frequency, when attain their zero value at different instants, the quantities are said to have a phase difference. This angle between zero points (and are becoming positive) of two alternating quantities is called *angle of phase difference.*

In Fig. 7.13, two alternating currents of magnitude I_{m1} and I_{m2} are shown vectorially. Both the vectors are rotating at same angular velocity of *w* radian per second. The zero values are obtained by the two currents at different instants. Therefore, they are said to have a phase difference of angle *f.*

In other words, **the phase difference** *may be defined as the angular displacement between the maximum positive value of the two alternating quantities having the same frequency.*

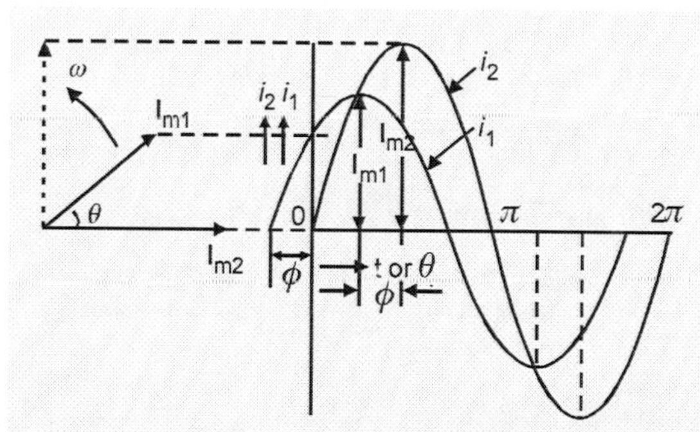

Fig. 7.13

The quantity which attains its +ve maximum value prior to the other is called a *leading* quantity, whereas, the quantity which attains its +ve maximum value after the other is called a *lagging* quantity. In this case current I_{ml} is leading current *w.r.t.* I_{m2} or is other words current I_{m2} is the lagging current *w.r.t.* I_{ml}.

7.16 ADDITION AND SUBTRACTION OF ALTERNATING QUANTITIES

In a.c. circuits, sometimes it is required to add or subtract the alternating quantities. In such cases, we proceed as follows:

Addition of alternating quantities

The given alternating quantities are represented as phasor and then they are added in the same manner as forces are added. Only phasors of the similar quantities are added *i.e.* either all the currents are added or all the voltages are added. Voltages and currents are never added with each other. For addition, usually the method of components is applied:

(*i*) **Method of components:** In this method each phasor is resolved into horizontal and vertical components. The horizontal components are added algebraically to obtain the resultant horizontal component I_{XX}. Similarly, vertical components are summed up algebraically to obtain the resultant vertical component I_{YY}.

Consider an a.c. parallel circuit consisting of three branches each carrying a current of i_1, i_2 and i_3 respectively as shown in Fig. 7.14. Let the three currents be represented by;

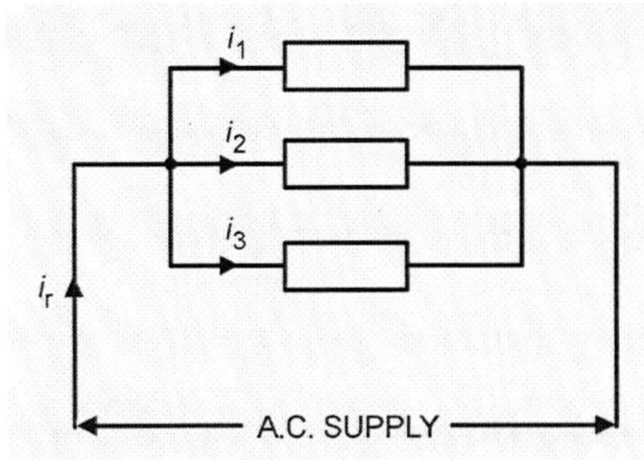

Fig. 7.14

$$i_1 = I_{m1} \sin(\omega t + \theta_1);$$

$$i_2 = I_{m2} \sin \omega t$$

and $$i_3 = I_{m3} \sin(\omega t - \theta_2)$$

The maximum values of the three currents I_{m1}, I_{m2} and I_{m3} are represented by the phasors as shown in Fig. 7.15 (*a*). Resolving the components horizontally and vertically.

Algebraic sum of horizontal components;

$$I_{XX} = I_{m1} \cos \theta_1 + I_{m2} + I_{m3} \cos \theta_2$$

Algebraic sum of vertical components;

$$I_{YY} = I_{m1} \sin \theta_1 + 0 - I_{m3} \sin \theta_2$$

Maximum value of resultant components;

$$I_{mr} = \sqrt{(I_{XX})^2 + (I_{YY})^2}$$

If ϕ is the phase difference (leading) between resultant current and horizontal axis as shown in Fig. 7.15 (*b*). Then

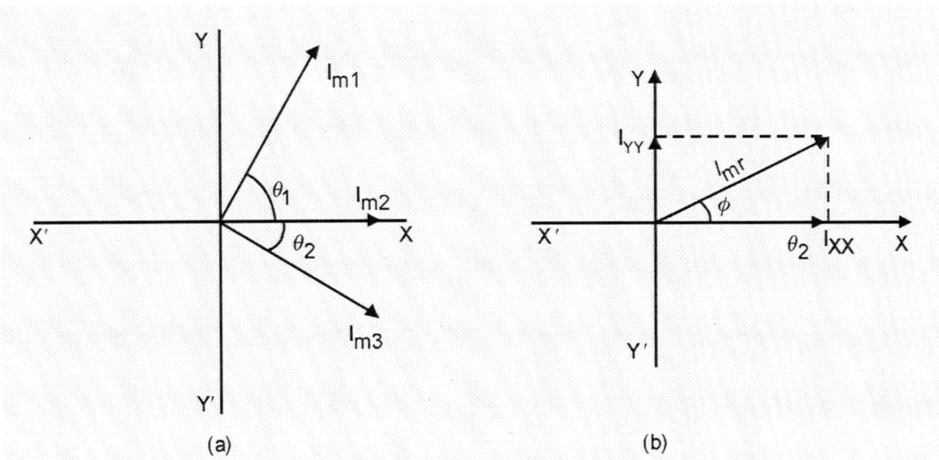

Fig. 7.15

$$\phi = \tan^{-1} \frac{I_{YY}}{I_{XX}}$$

The instantaneous value of the resultant current is given by the relation;

$$i_r = I_{mr} \sin(\omega t + \phi)$$

However, if I_{YY} comes out to be negative, the angle of phase difference will be lagging (*i.e.* $-\phi$). Then the instantaneous value of the resultant current will be given by the relation

$$i_r = I_{mr} \sin(\omega t - \phi)$$

Subtraction of alternating quantities

The methods explained above (*i.e.* parallelogram method and method of components) are also applied for the subtraction of an alternating quantity. The only difference is that in this case, the phasor of the alternating quantity which is to be subtracted is reversed or represented 180° out of phase. Then it is added with the other alternating quantity (or quantities) as usual.

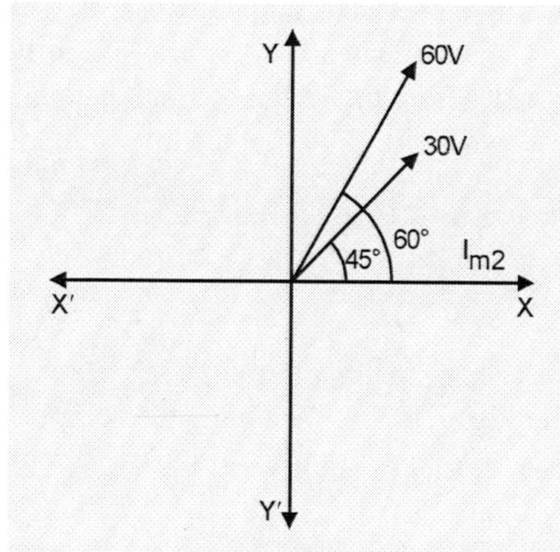

Fig. 7.16

Example 7.4. *Two ac voltage are represented by:*

$$\upsilon_1(t) = 30 \sin(314\,t + 45°);$$
$$\upsilon_2(t) = 60 \sin(314\,t + 60°)$$

Calculate the resultant $\upsilon(t)$ and express in the form:

$$\upsilon(t) = V_m \sin(314\,t \pm \phi)$$

Solution: $\upsilon_1(t) = 30 \sin(314\,t + 45°)$;

$\upsilon_2(t) = 60 \sin(314\,t + 60°)$ are shown in Fig. 7.16.

Resolving the vectors V_{1m} and V_{2m} horizontally and vertically we get,

$$\Sigma V_{xx} = 30 \cos 45° + 60 \cos 60°$$
$$= 30 \times \frac{1}{\sqrt{2}} + 60 \times \frac{1}{2} = 51.21$$
$$\Sigma V_{yy} = 30 \sin 45° + 60 \sin 60°$$
$$= 30 \times \frac{1}{\sqrt{2}} + 60 \times \frac{\sqrt{3}}{2} = 73.17$$

Maximum value of resultant voltage,

$$V_{r(m)} = \sqrt{(\Sigma V_{xx})^2 + (\Sigma V_{yy})^2}$$
$$= \sqrt{(51\cdot21)^2 + (73\cdot17)^2} = 89.31$$

$$\text{Phase angle, } \phi_r = \tan^{-1}\frac{\Sigma V_{yy}}{\Sigma V_{xx}} = \tan^{-1}\frac{73\cdot17}{51\cdot21} = 55°$$

Resultant voltage, $\upsilon(t) = V_{r(m)} \sin(\omega t + \phi_r) = 89.31 \sin(314\,t + 55°)$ (*Ans.*)

Example 7.5. *Draw a phasor diagram showing the following voltages.*

$v_1 = 100 \sin 500\,t$; $\quad v_2 = 200 \sin(500\,t + \pi/3)$

$v_3 = -50 \cos 500\,t$; $\quad v_4 = 150 \sin(500\,t - \pi/4)$

Find RMS value of resultant voltage.

Solution: $v_1 = 100 \sin 500\,t;$

$$v_2 = 200 \sin\left(500\,t + \frac{\pi}{3}\right)$$

$$v_3 = -50 \cos 500\,t = -50 \sin\left(\frac{\pi}{2} - 500\,t\right)$$

$$= 50 \sin (500\,t - \pi/2)$$

$$v_4 = 150 \sin (150\,t - \pi/4)$$

All the four voltages are shown vectorially in Fig. 7.17.

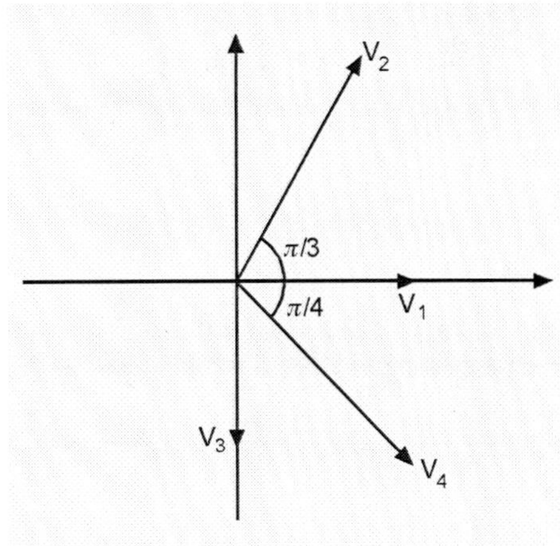

Fig. 7.17

Resolving the phasors in horizontal axis;

$$V_{xx} = V_1 \cos 0 + V_2 \cos \frac{\pi}{3} + V_3 \cos \frac{\pi}{2} + V_4 \cos \frac{\pi}{4}$$

$$= 100 \times 1 + 200 \times 0.5 + 50 \times 0 + 150 \times 0.707$$

$$= 306.05 \text{ V}$$

Resolving the phasors in vertical axis;

$$V_{yy} = V_1 \sin 0 + V_2 \sin \frac{\pi}{3} - V_3 \sin \frac{\pi}{2} - V_4 \sin \frac{\pi}{4}$$

$$= 100 \times 0 + 200 \times 0.866 - 50 \times 1 - 150 \times 0.707 = 17.15 \text{ V}$$

Maximum value of resultant voltage;

$$V_{mr} = \sqrt{V_{xx}^2 + V_{yy}^2} = \sqrt{(306 \cdot 05)^2 + (17 \cdot 15)^2} = 306.53 \text{ V}$$

RMS value of resultant voltage, $V_{rms(r)} = \frac{V_{mr}}{\sqrt{2}} = \frac{306 \cdot 53}{\sqrt{2}}$ = **216.75 V** (*Ans.*)

(*ii*) Frequency, $f = \frac{440}{2\pi}$ = **70 Hz** (*Ans.*)

R M S value of resultant voltage, $V_{rms(r)} = \frac{V_m}{\sqrt{2}} = \frac{22 \cdot 36}{\sqrt{2}}$ = **15.81 V** (*Ans.*)

Example 7.6. *Two currents i_1 and i_2 are given by the expressions: (i) 40 sin (314 t + π/6) (ii) 20 sin (314 t – π/3)*

Find $i_1 - i_2$ and express the answer in the same form as individual currents. Find also the r.m.s. value and frequency of the resultant current.

Solution: The two currents are represented by phasors as shown in Fig. 7.18. Since current i_2 is to be subtracted from i_1, reverse the phasor I_{m2} (dotted phasor) and find its vector sum with I_{m1}.

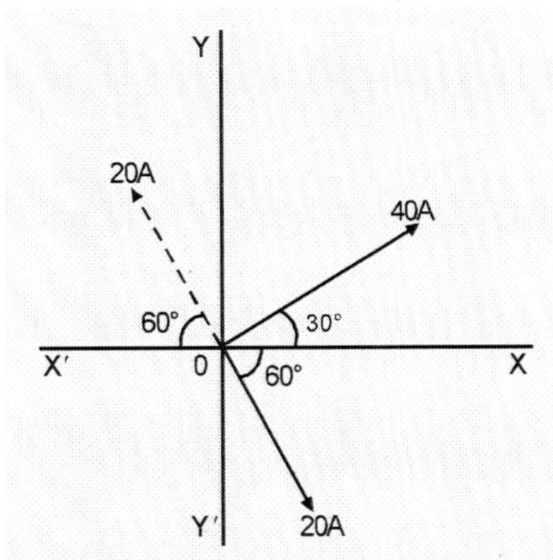

Fig. 7.18

Resolving the components along x-axis;

$$I_{XX} = 40 \cos 30° - 20 \cos 60°$$

$$= 40 \times 0.866 - 20 \times 0.5 = 24.64 \text{ A}$$

Resolving the components along y-axis;

$$I_{YY} = 40 \sin 30° + 20 \sin 60°$$
$$= 40 \times 0.5 + 20 \times 0.866 = 37.32 \text{ A}$$

Maximum value of resultant current,

$$I_{mr} = \sqrt{(I_{XX})^2 (I_{YY})^2} = \sqrt{(24 \cdot 64)^2 + (37 \cdot 32)^2}$$
$$= 44.72 \text{ A}$$

Let ϕ be the angle which resultant current makes with x-axis;

$$\tan \phi = \frac{I_{YY}}{I_{XX}} = \frac{37 \cdot 32}{24 \cdot 64} = 1.5146 \quad \therefore \ \phi = \tan^{-1} 1.5146 = 56.56°$$

$\therefore$ Expression for the resultant current,

$$i_r = I_{mr} \sin(\omega t + \phi) = 44.72 \sin(314\,t + 56.56°) \ (Ans.)$$

or $\quad i_r = 44.72 \sin(314\,t + 0.9872)$ [where ϕ is in radians]

R.M.S. value of resultant current,

$$I_r = \frac{44 \cdot 72}{\sqrt{2}} = 31.62 \ (Ans.)$$

From the equation, $\omega = 314$ or $2\pi f = 314$

$\therefore$ Frequency, $f = 314/2\pi = 50$ Hz (*Ans.*)

TEST YOUR LEARNING-2

Type-I: Short Answer Questions

1. How will you define phase of an alternating quantity?
2. What do you mean by phase difference?
3. What do you mean by a leading quantity in an ac system?
4. What do you mean by a lagging quantity in an ac system?
5. How are the alternating quantities added or subtracted?

Type-II: Numerical for Practice

1. Three voltages represented by $e_1 = 10 \sin \omega t$, $e_2 = 15 \sin(\omega t + p/4)$ and $e_3 = 20 \cos(\omega t - p/6)$ act together in a circuit. Find an expression for the resulting voltage.

   (*Ans.* $41.433 \sin(\omega t + 42.38°)$)

2. Represent the following alternating currents as phasors on the same axes and determine phase angle of all other currents with I_1: $i_1 = 10 \sin(\omega t + \pi/3)$; $i_2 = 20 \cos(\omega t - \pi/6)$; $i_3 = 15 \cos(\omega t - \pi/3)$; $i_4 = 12 \sin(\omega t - \pi/6)$.

   (*Ans.* 60°; 30°; 90°)

3. Two currents i_1 and i_2 are given by the expression: $i_1 = 15 \sin(\omega t + \pi/3)$ and $i_2 = 5 \sin(\omega t - \pi/3)$ Find $i_1 - i_2$ and express the answer in the same form as individual currents.

(*Ans.* $18.028 \sin(\omega t + 73.89°)$)

4. *The instantaneous values of two alternating voltages are represented by* $\upsilon_1 = 60 \sin\theta$ *and* $\upsilon_2 = 40 \sin(\theta - \pi/3)$. *Derive expression for the instantaneous values of* (*i*) *the sum and* (*ii*) *the difference of these voltages.* (*Ans.* $87.2 \sin(\theta - 0.13\pi)$ *volt*; $52.9 \sin(\theta + 0.227\pi)$ **volt**)

Type-III: Test Questions

1. What is phase and phase difference? Explain with the help of phasor and wave diagrams?
2. Explain the method of components used for the addition and subtraction of alternating quantities.

7.17 A.C. CIRCUITS

Alternating supply is invariably used for domestic and industrial applications. *The path for the flow of alternating current is called an* **a.c. circuit**. In d.c. circuits, the opposition to the flow of current is only resistance of the circuit. Whereas, in a.c. circuits, the opposition to the flow of current is due to resistance (R), inductive reactance ($X_L = 2\pi f L$) and capacitive reactance ($X_C = 1/2\pi f C$) of the circuit.

7.18 A.C. CIRCUIT CONTAINING RESISTANCE ONLY

The circuit containing a pure resistance of R ohms is shown in Fig. 7.19.

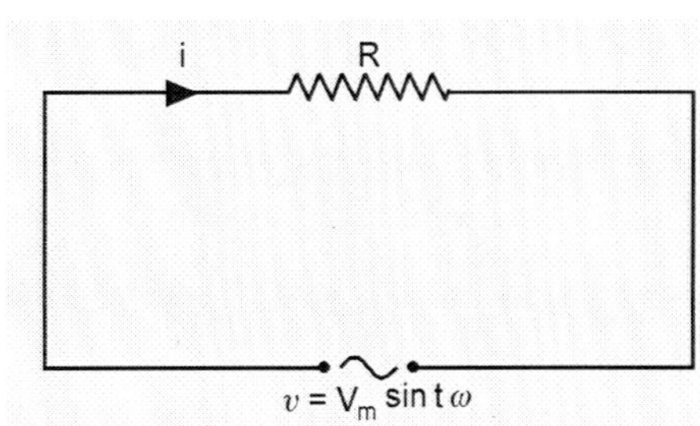

Fig. 7.19

Let the alternating voltage applied across the circuit be given by the equation;

$$v = V_m \sin \omega t \quad ...(i)$$

Then the instantaneous value of current flowing through the resistor will be;

$$i = \frac{v}{R} = \frac{V_m}{R} \sin \omega t \quad ...(ii)$$

The value of current will be maximum when

$$\omega t = 90° \text{ or } \sin \omega t = 1$$

$$I_m = V_m/R$$

Substituting this value is equation (*ii*), we get,

$$i = I_m \sin \omega t \quad ...(iii)$$

Phase angle: From equation (*i*) and (*iii*), it is clear that there is no phase difference between applied voltage and current flowing through pure resistive circuit *i.e.* phase angle between voltage and current is zero. The phasor diagram and wave diagram is shown in Fig. 7.20 (*a*) and (*b*) respectively.

Hence, in an a.c. circuit containing pure resistance, **current is in phase with the voltage.**

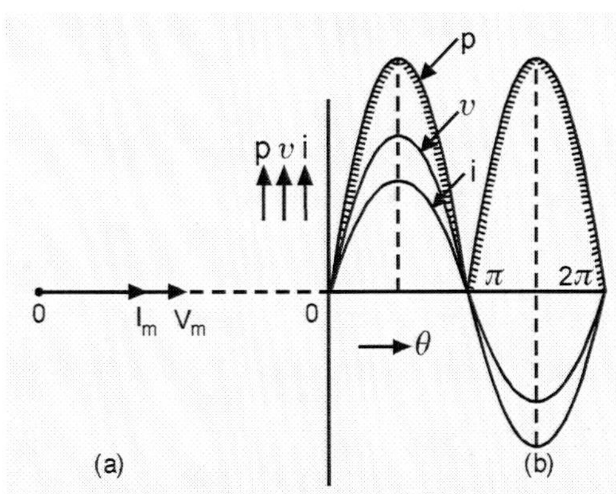

Fig. 7.20

Power

Instantaneous power, $p = vi = (V_m \sin \omega t)(I_m \sin \omega t)$

$$= \frac{V_m I_m}{2} 2 \sin^2 \omega t$$

$$= \frac{V_m}{\sqrt{2}} \frac{I_m}{\sqrt{2}} (1 - \cos 2\omega t) = \frac{V_m}{\sqrt{2}} \frac{I_m}{\sqrt{2}} - \frac{V_m}{\sqrt{2}} \frac{I_m}{\sqrt{2}} \cos 2\omega t$$

Average power consumed in the circuit over a complete cycle.

$$P = \text{average of } \frac{V_m}{\sqrt{2}} \frac{I_m}{\sqrt{2}} - \text{average of } \frac{V_m}{\sqrt{2}} \frac{I_m}{\sqrt{2}} \cos \omega t$$

or $\quad P = V_{r.m.s.} I_{r.m.s.} - \text{zero} \quad$ or $\quad P = VI$

Power curve: Fig. 7.20. (*b*) shows the power curve for a pure resistive circuit. Points on the power curve are obtained from the product of the corresponding instantaneous values of voltage and current.

7.19 A.C. CIRCUIT CONTAINING PURE INDUCTANCE ONLY

The circuit containing a pure inductance of L henry is shown in Fig. 7.21.

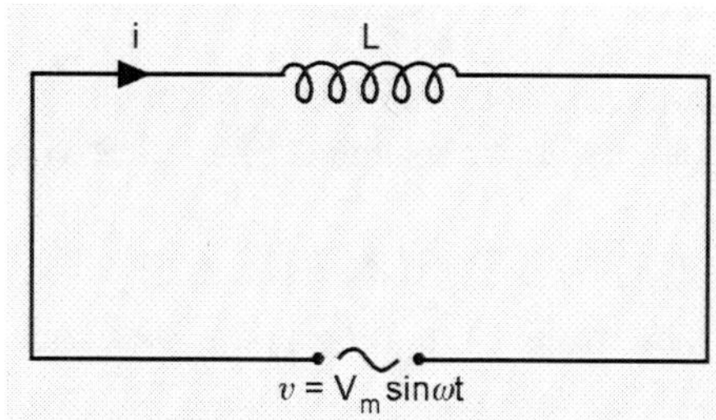

Fig. 7.21

Let the alternating voltage applied across the circuit be given by the equation;

$$v = V_m \sin \omega t \qquad ...(i)$$

As a result, an alternating current i flows through the inductance which induces an e.m.f. in it, given by the relation;

$$e = -L \frac{di}{dt}$$

This induced e.m.f. is equal and opposite to the applied voltage.

$$\therefore \quad v = -e = -\left(-L \frac{di}{dt}\right)$$

or $\quad V_m \sin \omega t = L \frac{di}{dt} \quad$ or $\quad di = \frac{V_m}{L} \sin \omega t \, dt$

Integrating both sides,

$$\int di = \int \frac{V_m}{L} \sin \omega t \, dt \quad \text{or} \quad i = \frac{V_m}{\omega L}(-\cos \omega t)$$

or $$i = \frac{V_m}{\omega L} \sin(\omega t - \pi/2) = \frac{V_m}{X_L} \sin(\omega t - \pi/2) \qquad ...(ii)$$

where $X_L = \omega L$ is the opposition offered to the flow of alternating current by a pure inductance and is called **inductive reactance.**

The value of current will be maximum when $\sin(\omega t - \pi/2) = 1$ *i.e.* $I_m = \frac{V_m}{X_L}$

$$\therefore \quad i = I_m \sin(\omega t - \pi/2) \qquad ...(iii)$$

Phase angle: From equation (*i*) and (*iii*), it is clear that current flowing through a pure inductive circuit lags behind the applied voltage v by 90°. The phasor diagram is shown in Fig. 7.22 (*a*) and 7.22 (*b*) respectively.

Hence, in an a.c. circuit containing pure inductance, **current lags behind the voltage by 90°.**

Power: Instantaneous power, $p = vi = V_m \sin \omega t \times I_m \sin (\omega t - \pi/2)$

$$= V_m I_m \sin \omega t \cos \omega t = \frac{V_m I_m}{2} 2 \sin \omega t \cos \omega t$$

$$= \frac{V_m}{\sqrt{2}} \frac{I_m}{\sqrt{2}} \sin 2\omega t$$

Average power consumed in the circuit over a complete cycle,

$$P = \text{average } \frac{V_m}{\sqrt{2}} \cdot \frac{I_m}{\sqrt{2}} \sin 2\omega t = \text{zero}$$

Hence, average power consumed in a pure inductive circuit is **zero.**

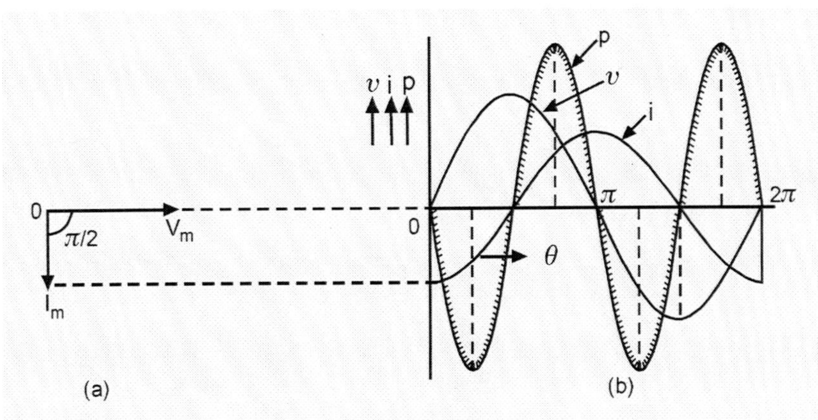

Fig. 7.22

Power curve: The power curve for a pure inductive circuit is shown in Fig. 7.22 (*b*). It is very clear that average power in a half cycle (one alternation) is zero, as the negative and positive loop area under power curve is the same.

It is interesting to note that during the first quarter cycle, what so ever power (or energy) is supplied by the source to the inductance (or coil) is stored in the magnetic field set-up around it. However, in the next quarter cycle, the magnetic field collapses and the power (or energy) stored in the field is returned to the source. This process is repeated in each and every alternation. **Hence, no power or energy is consumed in this circuit.**

7.20 A.C. CIRCUIT CONTAINING PURE CAPACITOR ONLY

The circuit containing a pure capacitor of capacitance C farad is shown in Fig. 7.23.

Let the alternating voltage applied across the circuit be given by the equation;

$$v = V_m \sin \omega t \quad \text{...(i)}$$

Charge on the capacitor at any instant,

$$q = C v$$

Current flowing through the circuit,

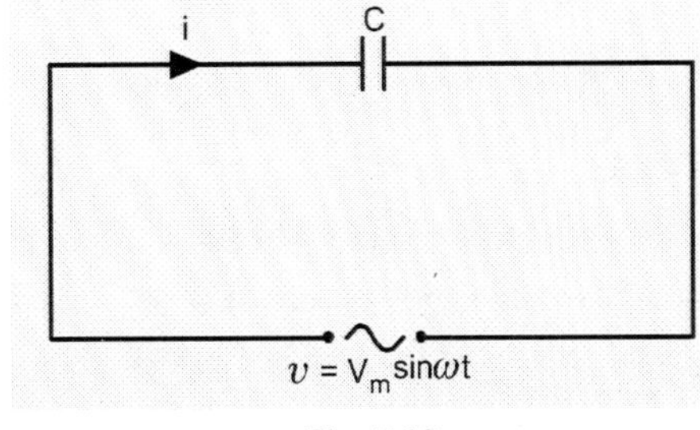

Fig. 7.23

$$i = \frac{d}{dt}q = \frac{d}{dt}(Cv)$$

or $$i = \frac{d}{dt}C\,V_m \sin \omega t = C\,V_m \frac{d}{dt} \sin \omega t$$

or $$i = \omega C\,V_m \cos \omega t = \frac{V_m}{1/\omega C} \sin(\omega t + \pi/2) = \frac{V_m}{X_C} \sin(\omega t + \pi/2) \qquad ...(ii)$$

where $X_C = 1/\omega C$ is the opposition offered to the flow of alternating current by a pure capacitor and is called *capacitive reactance.*

The value of current will be maximum when $\sin(\omega t + \pi/2) = 1$

i.e. $I_m = V_m/X_C$

Substituting this value is equation (*ii*), we get,

$$i = I_m \sin(\omega t + \pi/2) \qquad ...(iii)$$

Phase angle: From equation (*i*) and (*iii*), it is clear that current flowing through pure capacitive circuit leads the applied voltage by 90°. the phasor diagram and wave diagram is shown in Fig. 7.24 (*a*) and 7.24 (*b*) respectively.

Hence, in an a.c. circuit containing pure capacitance **current leads the voltage by 90°.**

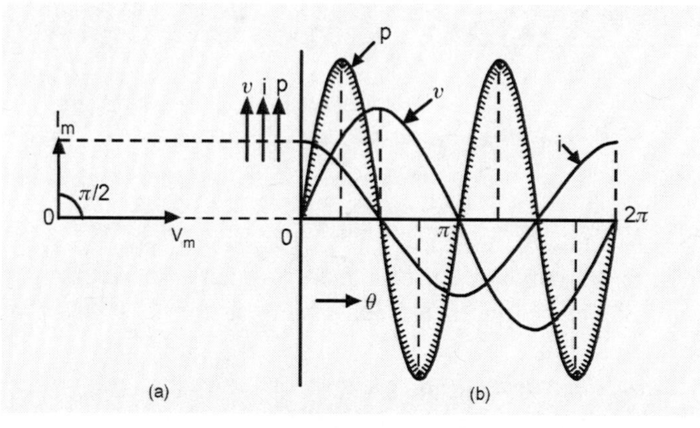

Fig. 7.24

Power: Instantaneous power, $p = v\,i = V_m \sin \omega t \times I_m \sin(\omega t + \pi/2)$

$$= V_m I_m \sin \omega t \cos \omega t = \frac{V_m}{\sqrt{2}} \frac{I_m}{\sqrt{2}} \sin 2\,\omega t$$

or Average power over a complete cycle,

$$P = \text{zero}$$

Hence, average power consumed in a pure capacitive circuit is **zero.**

Power curve: The power curve for a pure capacitive circuit is shown in Fig. 7.24(*b*). It is very clear from the curve that average power in a half cycle (one alternation) is zero since the positive and negative loop area under power curve is the same.

It is interesting to note that, during the first quarter cycle, what so ever power (or energy) is supplied by the source to the capacitor is stored in the electric field set-up between the capacitor plates. In the next quarter cycle, the electric field collapses and the power (or energy) stored in the field is returned to the source. This process is repeated in each alternation. **Hence, no power is consumed by this circuit.**

Example 7.7. *An a.c. circuit consists of a pure resistance of 10 ohm and is connected across an a.c. supply of 230 V, 50 Hz. Calculate (i) current; (ii) power consumed and (iii) write down the equation for voltage and current.*

Solution: (*i*) Current in the circuit, $I = \frac{V}{R} = \frac{230}{10} = 23$ A (*Ans.*)

(*ii*) Power consumed, $P = VI = 230 \times 23 = 5290$ W (*Ans.*)

(*iii*) Maximum value of applied voltage, $V_m = \sqrt{2}\,V = \sqrt{2} \times 230 = 325.27$ V

Maximum value of current, $I_m = \sqrt{2} \times 23 = 32.53$ A

Angular velocity, $\omega = 2\pi f = 2\pi \times 50 = 314.16$ rad/s

Equation for applied voltage;

$$v = V_m \sin \omega t = 325.27 \sin 314.16\,t \text{ (Ans.)}$$

As in a pure resistive circuit, voltage and current are in phase with each other, therefore, current is given by the equation;

$$i = I_m \sin \omega t = 32.53 \sin 314.16\,t \text{ (Ans.)}$$

Example 7.8. *An inductive coil having negligible resistance and 0.1 henry inductance is connected across 200 V, 50 Hz supply. Find (i) the inductive reactance (ii) r.m.s. value of current (iii) power and (iv) equations for voltage and current.*

Solution: Inductive reactance, $X_L = 2\pi L = 2\pi \times 50 \times 0.1 = 31.416\ \Omega$ (*Ans.*)

Current, $I = V/X_L = 200/31.416 = 6.366$ A (*Ans.*)

Power, P = **Zero** (*Ans.*)

Now, $V_m = \sqrt{2}\,V = \sqrt{2} \times 200 = 282.84$ V;

$I_m = \sqrt{2}\,I = \sqrt{2} \times 6.366 = 9$ A

and $\omega = 2\pi f = 314$ rad/s

$\therefore$ $v = V_m \sin \omega t = 282.84 \sin 314\,t$ (*Ans.*)

In pure inductive circuit, current lags behind voltage by $\pi/2$ radian.

$\therefore$ $i = I_m \sin(\omega t - \pi/2) = 9 \sin(314\,t - \pi/2)$ (*Ans.*)

Example 7.9. *A capacitor has a capacitance of 30 microfarad. Find its capacitive reactance for frequencies of 25 and 50 Hz. Find in each case the current if the supply voltage is 440 V.*

Solution: Capacitance of the capacitor, $C = 30 \times 10^{-6}$ F

Supply voltage, $V = 440$ V

When supply frequency, $f_1 = 25$ Hz

Capacitive reactance, $X_{C1} = \frac{1}{\omega_1 C} = \frac{1}{2\pi f_1 C} = \frac{1}{2\pi \times 25 \times 30 \times 10^{-6}} = 212.2\ \Omega$ (*Ans.*)

Current in the circuit, $I_1 = \frac{V}{X_{C1}} = \frac{440}{212 \cdot 2} = 2.073$ A (*Ans.*)

When supply frequency, $f_2 = 50$ Hz

Capacitive reactance, $X_{C2} = \frac{1}{\omega_2 C} = \frac{1}{2\pi f_2 C} = \frac{1}{2\pi \times 50 \times 30 \times 10^{-6}} = 106.1\ \Omega$ (*Ans.*)

Current in the circuit, $I_2 = \frac{V}{X_{C2}} = \frac{440}{106 \cdot 1} = 4.146$ A (*Ans.*)

TEST YOUR LEARNING-2

Type-I: Short Answer Questions

1. What is the phase difference between voltage and current when sinusoidal ac voltage is applied across a pure resistor?
2. What is the phase difference between V and I in pure inductive circuit.
3. What is the power consumption in a pure capacitive circuit?

Type-II: Numerical for Practice

1. An a.c. circuit consists of a pure resistance of 20 ohm and is connected across an a.c. supply of 240 V, 50 Hz. Calculate (i) current; (ii) power consumed and (iii) write down the equation for voltage and current. (*Ans.* 12A, 2880 W, $v = 339.4 \sin 314\, t$)
2. An inductive coil having negligible resistance and 0.1 henry inductance is connected across 230 V, 50 Hz supply. Find (i) the inductive reactance (ii) r.m.s. value of current (iii) power and (iv) equations for voltage and current.
(*Ans.* 31.416 Ω, 7.32 A, zero, $v = 325.27 \sin 314\, t$, $i = 10.35 \sin (314t - \pi/2)$)
3. A capacitor has a capacitance of 20 microfarad. Find its capacitive reactance for frequencies of 50 Hz. Find the capacitive reactance and current if the supply voltage is 230 V.
(*Ans.* 159.15 Ω, 1.445 A)

Type-III: Test Questions

1. When sinusoidal ac voltage is applied across a pure inductor, show that power consumed in the circuit is zero. Also draw the phasor and wave diagram for voltage and current.
2. Show that in a pure capacitive ac circuit, power consumption is zero.

7.21 A.C. SERIES CIRCUITS

In actual practice, a.c. circuits contain two or more than two such components connected in series or parallel. A series circuit is a circuit in which each component carries the same current. An a.c. series circuit may be;

(*i*) $R - L$ series circuit (*ii*) $R - C$ series circuit (*iii*) $R - L - C$ series circuit.

7.22 R – L SERIES CIRCUIT

A circuit that contains a pure resistance R ohms connected in series with a coil having pure inductance of L henry is known as $R - L$ series circuit. This is the most general case that we come across in practice.

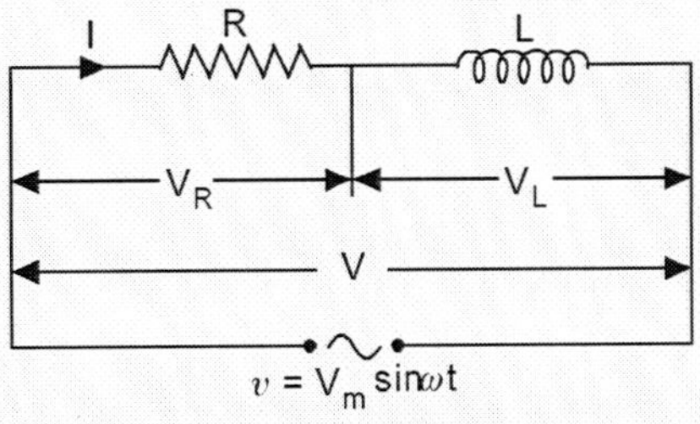

Fig. 7.25

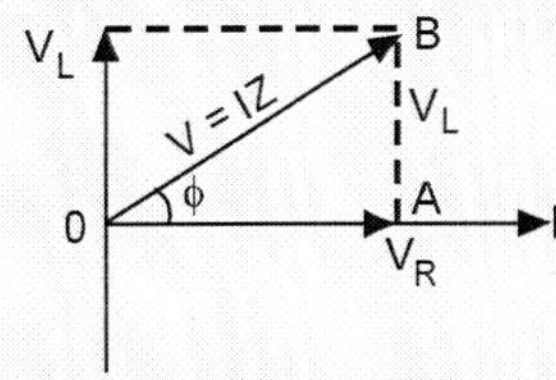

Fig. 7.26

An $R-L$ series circuit and its phasor diagram is shown in Fig. 7.25 and 7.26 respectively. To draw the phasor diagram, current I (r.m.s. value) is taken as the reference vector. Voltage drop in resistance V_R ($= IR$) is taken in phase with current vector, whereas, voltage drop in inductive reactance V_L ($= IX_L$) is taken 90° ahead of the current vector (since current lags behind the voltage by 90° in pure inductive circuit). The vector sum of these two voltages (drops) is equal to the applied voltage V (r.m.s. value).

Now, $V_R = IR$ and $V_L = IX_L$ (where $X_L = 2\,\pi f L$)

In right angle triangle OAB

$$V = \sqrt{(V_R)^2 + (V_L)^2} = \sqrt{(IR)^2 + (IX_L)^2} = I\sqrt{R^2 + X_L^2}$$

or
$$I = \frac{V}{\sqrt{R^2 + X_L^2}} = \frac{V}{Z}$$

where $Z = \sqrt{R^2 + X_L^2}$ is the total opposition offered to the flow of alternating current by an $R - L$ series circuit and is called *impedance* of the circuit. It is measured in ohm.

Phase angle

From the phasor diagram shown in Fig. 7.26, it is clear that current, in this circuit, lags behind the applied voltage by an angle *f*, called *phase angle*.

From phasor diagram; $\tan\phi = \dfrac{V_L}{V_R} = \dfrac{IX_L}{IR} = \dfrac{X_L}{R}$ or $\phi = \tan^{-1} X_L/R$

Power: If the alternating voltage applied across the circuit is given by the equation.

$$v = V_m \sin \omega t$$

Then, $$i = I_m \sin(\omega t - \phi)$$

∴ Instantaneous power, $p = v\,i$

$$= V_m \sin\omega t . I_m \sin(\omega t - \phi) = \frac{V_m I_m}{2}\, 2\sin\omega t \sin(\omega t - \phi)$$

$$= \frac{V_m}{\sqrt{2}}\frac{I_m}{\sqrt{2}}[\cos\phi - \cos(2\omega t - \phi)]$$

$$= \frac{V_m}{\sqrt{2}}\frac{I_m}{\sqrt{2}}\cos\phi - \frac{V_m}{\sqrt{2}}\frac{I_m}{\sqrt{2}}\cos(2\omega t - \phi)$$

Power curve for $R - L$ series circuit is also shown in Fig. 7.27 (*b*).

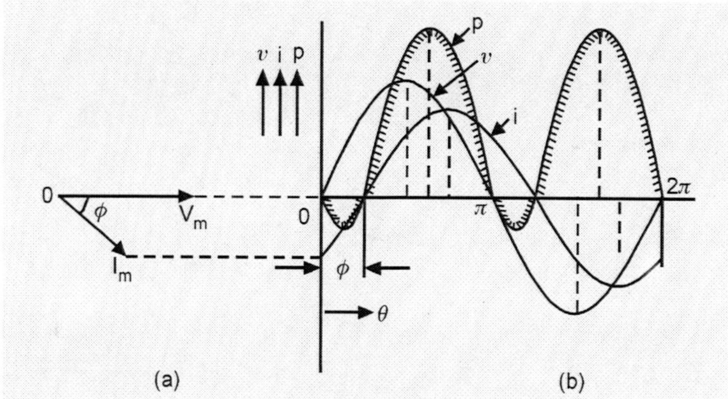

Fig. 7.27

Average power consumed in the circuit over a complete cycle,

$$P = \text{average of } \frac{V_m}{\sqrt{2}} \frac{I_m}{\sqrt{2}} \cos\phi - \text{average of } \frac{V_m}{\sqrt{2}} \frac{I_m}{\sqrt{2}} \cos(2\omega t - \phi)$$

or $$P = \frac{V_m}{\sqrt{2}} \frac{I_m}{\sqrt{2}} \cos\phi - \text{zero} = V_{r.m.s.} I_{r.m.s.} \cos\phi = VI\cos\phi$$

where $\cos\phi$ is called *power factor* of the circuit.

From phasor diagram; $\cos\phi = \frac{V_R}{V} = \frac{IR}{IZ} = \frac{R}{Z}$

Thus, **power factor** *is defined as the cosine of the angle between voltage and current in an a.c. circuit. It may also be defined as the ratio of resistance to impedence of an a.c. circuit.*

Alternatively: Power, $P = VI\cos\phi = IZ.I.\frac{R}{Z} = I^2 R$

This shows that power is actually consumed in resistance only; inductance does not consume any power.

Power curve: The phasor diagram and wave diagram for voltage and current are shown in Fig. 7.27 (*a*) and (*b*) respectively where applied voltage ($v = V_m \sin \omega t$) is taken as reference quantity. The power

7.23 IMPEDENCE TRIANGLE

The simplified phasor diagram of $R - L$ series circuit is shown in Fig. 7.28. When each side of this phasor diagram is divided by a common factor I, we get another right angled triangle, as shown in Fig. 7.29, whose sides represent R, X_L and Z. Such a triangle is known as *impedance triangle*.

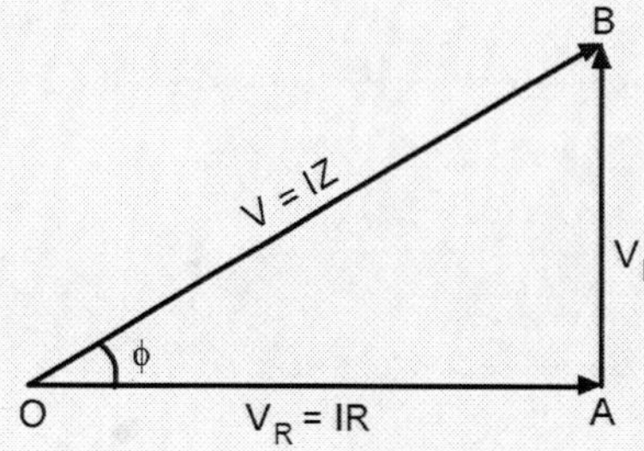

Fig. 7.28

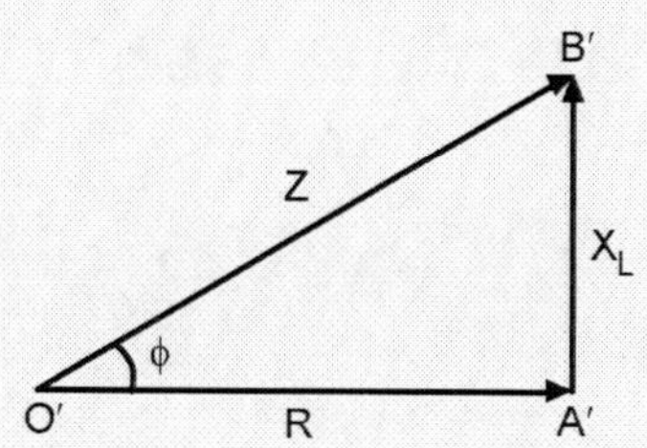

Fig. 7.29

Thus, a right angled triangle whose base represents circuit resistance, perpendicular represents circuit reactance and hypotenuse represents circuit impedance is called an **impedance triangle.**

The concept of impedance triangle is useful since it enables us to calculate:

(*i*) the impedance of the circuit, $Z = \sqrt{R^2 + X^2}$

(*ii*) the power factor of the circuit, $\cos \phi = R/Z$

(*iii*) phase angle, $\phi = \tan^{-1} X_L/R$.

7.24 POWER FACTOR AND ITS IMPORTANCE

In a.c. circuits, the power factor may be expressed as:

$$p.f. = \cos \phi = R/Z = \text{true power/ apparent power}$$

In case of pure resistive circuit, current is in phase with circuit voltage *i.e.* $\phi = 0$. Therefore, power factor of the circuit, $\cos \phi = 1$. Whereas, in case of pure inductive or capacitive circuit, current is 90° out of phase with circuit voltage *i.e.* $\phi = 90°$. Therefore, power factor of the circuit $\cos \phi = 0$. For circuits having resistance-inductance; resistance-capacitance or resistance-inductance and capacitance, the power factor lies between 0 and 1. *It may be noted that the value of p.f. can never be* **more than one.**

Usually, the word *lagging* or *leading* is attached with the numerical value of p.f. to signify whether the current lags or leads the voltage. In inductive circuits, current always lags behind the voltage and their power factors are mentioned as lagging p.f. Whereas, for capacitive circuits, the power factor is mentioned as leading p.f. since in these cases current always leads the voltage vector.

Importance of Power Factor

The power factor of an a.c. circuit plays an important role in the power system. Since power of an a.c. circuit is given by the relation;

$$P = VI \cos \phi = \text{or } I = \frac{P}{V \cos \phi}$$

From the above relation, it is clear that for fixed power at constant voltage, the current drawn by the circuit increases with decrease in p.f.

Thus at low p.f., a.c. circuits draw more current from their mains and results in the following disadvantages:

1. **Greater conductor size:** At low p.f., the conductors are to carry more current for the same ower therefore, they require larger area of cross-section.
2. **Poor efficiency:** At low power factors, the conductors have to carry larger current which increases copper losses (I^2R) and results in poor efficiency.
3. **Larger voltage drop:** At lower power factors, the conductors have to carry larger current which increases voltage drop (IR) in the system and results in poor regulation.
4. **Larger kVA rating of equipment:** The kVA rating of electrical machines and equipment connected in the power system such as alternators, transformers, switch gears etc. will be more at lower power factors since it is inversely proportional to power factor (*i.e.* kVA = kW/ cos *f*).

To improve the power factor of an a.c. circuit **a capacitor** *is connected across the circuit i.e. parallel to the circuit.*

Example 7.10. *A coil having a resistance of 12 Ω and an inductance of 0.1 H is connected across a 100 V, 50 Hz supply. Calculate (i) the reactance and impedance of the coil, (ii) the current, (iii) the phase difference between current and the applied voltage and (iv) the power factor. Draw also the phasor diagram showing voltage and current.*

Solution: The circuit is shown in Fig. 7.30.

(*i*) Reactance, $X_L = 2\,\pi f L = 2\,\pi\,50 \times 0.1 =$ **31.416 Ω** (*Ans.*)

Impedance, $Z = \sqrt{R^2 + X_L^2} = \sqrt{(12)^2 + (31\cdot416)^2} =$ **33.63 Ω** (*Ans.*)

(*ii*) Current, $I = \dfrac{V}{Z} = \dfrac{100}{33\cdot67} =$ **2.97 A** (*Ans.*)

(*iii*) Phase difference, $\phi = \tan^{1} \dfrac{X_L}{R} = \tan^{-1} \dfrac{31\cdot416}{12} = \tan^{-1} 2.618 =$ **69.1°** (*Ans.*)

(*iv*) Power factor, $\cos \phi =$ **0.3568 lag** (*Ans.*)

The phasor diagram for the circuit is shown in Fig. 7.31.

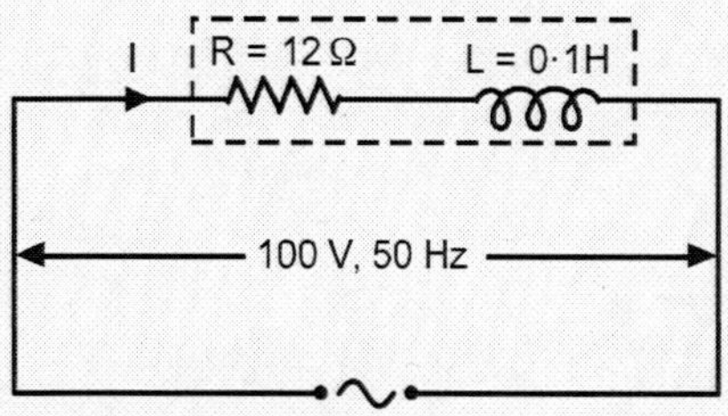

Fig. 7.30

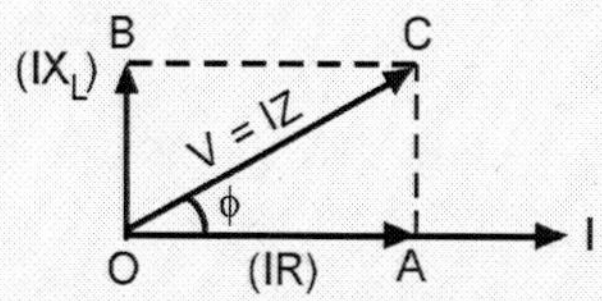

Fig. 7.31

Example 7.11. *The voltage and current through a circuit element are*

$$v = 50\,sin\,(314\,t + 55°)\ volt$$
$$i = 10\,sin\,(314\,t + 325°)\ ampere$$

Find the value of power drawn by the element.

Solution: Given $v = 50 \sin (314\,t + 55°)$ V

$$i = 10 \sin (314\,t + 325°)\ \text{A}$$

or $i = 10 \sin (314\,t - 35°)$ A

Now their phasor representation is shown in Fig. 7.32

∴ Phase difference between voltage and current is 90°.

Now power drawn by the circuit, $P = VI \cos \phi$

$$= \frac{50}{\sqrt{2}} \times \frac{10}{\sqrt{2}} \times \cos 90°$$

$$= \frac{500}{4} \times 0° = \textbf{0 W}\ (\textit{Ans.})$$

which indicates that element is pure inductive.

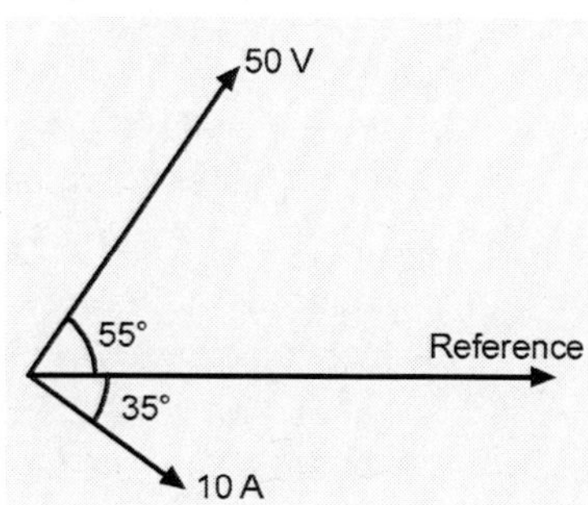

Fig. 7.32

Example 7.12. *A coil connected to* 100 V *d.c. supply draws* 10 *and the same coil when connected to* 100 V, *a.c. voltage of frequency* 50 *Hz draws* 5*A. Calculate the parameters of the coil and power factor.*

Solution: Let the resistance and inductance of the coil be R ohm and L henry respectively.

When coil in connected to *d.c.* supply, the opposition is only resistance of the coil,

$$\therefore \quad \text{Resistance of the coil, } R = \frac{V_{d.c.}}{I_{d.c.}} = \frac{100}{10} = \mathbf{10\Omega}\ (Ans.)$$

When coil is connected across *a.c.* supply of 100 V, 50 Hz, the opposition is impedance of the coil,

$$\therefore \quad \text{Impedance of the coil, } Z = \frac{V_{ac}}{I_{ac}} = \frac{100}{5} = 20\ \Omega$$

$$\text{Now, } Z = \sqrt{R^2 + X_L^2} \quad \text{or} \quad Z^2 = R^2 + X_L^2$$

$$\text{or} \quad X_L = \sqrt{Z^2 - R^2} = \sqrt{(20)^2 - (10)^2} = \sqrt{300}$$

$$\text{and} \quad L = \frac{X_L}{2\pi f} = \frac{\sqrt{300}}{2\pi \times 50} = \mathbf{55.13\ mH}\ (Ans.)$$

$$\therefore \quad \text{Parameters are } R = 10\ \Omega \text{ and } L = 55.13 \text{ mA}$$

$$\text{Power factor, } \cos\phi = \frac{R}{Z} = \frac{10}{20} = \mathbf{0.5\ lagging}\ (Ans.)$$

Example 7.13. *A non-inductive resistance of* 10 *ohm is connected in series with an inductive coil across* 200 *V,* 50 *Hz ac supply. The current drawn by the series combination is 10 ampere. The resistance of the coil is* 2 *ohm. Determine* (*i*) *inductance of the coil* (*ii*) *Power factor* (*iii*) *voltage across the coil.*

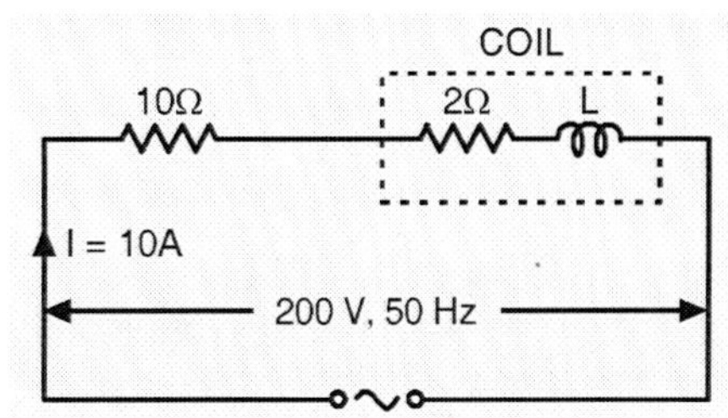

Fig. 7.33

Solution: Total impedance of the circuit, $Z = \frac{V}{I} = \frac{200}{10} = 20\ \Omega$

$$\text{Total resistance of the circuit, } R = 10 + 2 = 12\ \Omega$$

$$\text{Inductive reactance of the coil, } X_L = \sqrt{Z^2 - R^2} = \sqrt{(20)^2 - (12)^3} = 16\ \Omega$$

$$\text{or} \quad 2\pi f L = 16 \text{ or } L = \frac{16}{100\ \pi} = 50 \cdot 93 \text{ mH}$$

$$\therefore \quad \text{Inductance of the coil, } L = \mathbf{50.93\ mH}\ (Ans.)$$

$$(ii) \quad \text{Power factor of the circuit, } \cos\phi = \frac{R}{Z} = \frac{12}{20} = \mathbf{0.6\ lag}\ (Ans.)$$

$$(iii) \quad \text{Impedance of the coil, } Z_c = \sqrt{R_c^2 + X_L^2} = \sqrt{(2)^2 + (16)^2} = 16.124\ \Omega$$

$$\text{Voltage across the coil, } V_c = I Z_c = 10 \times 16.124 = \mathbf{161.24\ V}\ (Ans.)$$

$$\text{Power factor of the coil, } \cos\phi_c = \frac{R_C}{Z_c} = \frac{2}{16 \cdot 124} = 0.124 \text{ lagging}$$

Example 7.14. *A coil when connected to 200 V d.c. supply dissipates 2000 watt of power. When connected across 200 V a.c. supply of frequency 50 c.p.s., it dissipates 800 watt of power. Calculate the value of resistance and inductance of the coil.*

Solution: When d.c. is applied across the coil, the opposition is only resistance of the coil,

$$\therefore\ R = \frac{V_{dc}^2}{P_{dc}} = \frac{(200)^2}{2000} = \mathbf{20\ \Omega}\ (Ans.)$$

When a.c. is applied across the coil;

$$P_{ac} = I_{ac}^2 R \text{ or } I_{ac} = \sqrt{P_{ac}/R} = \sqrt{800/20} = 6.324 \text{ A}$$

$$\text{Impedance, } Z = \frac{V_{ac}}{I_{ac}} = \frac{200}{6\cdot 324} = 31.623\ \Omega$$

$$X_L = \sqrt{Z^2 - R^2} = \sqrt{(31\cdot 623)^2 - (20)^2} = 24.495\ \Omega$$

$$\text{Inductance, } L = \frac{X_L}{2\pi f} = \frac{24\cdot 495}{2\pi \times 50} = \mathbf{0.078\ H}\ (Ans.)$$

TEST YOUR LEARNING-3

Type-I: Short Answer Questions

1. What do you mean by impedance of an *RL* series circuit?
2. Draw an impedance triangle for an *RL* series circuit and mention its sides.
3. Define power factor for an *RL* series circuit.
4. State the disadvantages of poor power factor.
5. What do you mean by Q-factor of an *RL* circuit?

Type-II: Numerical for Practice

1. A coil having a resistance of 24 Ω and an inductance of 0.2 H is connected across a 200 V, 50 Hz supply. Calculate (i) the reactance and impedance of the coil, (ii) the current, (iii) the phase difference between current and the applied voltage and (iv) the power factor. Draw also the phasor diagram showing voltage and current. (***Ans.*** 67.26 Ω, 2.97 A, 69.1°, 0.3568 lag)
2. The voltage and current through a circuit element are

   $$v = 120 \sin(314\,t + 45°) \text{ volt}$$

   $$i = 12 \sin(314\,t + 315°) \text{ ampere}$$

   Find the value of power drawn by the element. (***Ans.*** zero)
3. A coil connected to 120 V d.c. supply draws 12 and the same coil when connected to 100 V, a.c. voltage of frequency 50 Hz draws 6A. Calculate the parameters of the coil and power factor. (***Ans.*** 10 Ω, 55.13 H, 0.5 lag)

4. A non-inductive resistance of 10 ohm is connected in seris with an inductive coil across 240 V, 50 Hz ac supply. The current drawn by the series combination is 12 ampere. The resistance of the coil is 2 ohm. Determine (i) inductance of the coil (ii) Power factor (iii) voltage across the coil. (*Ans.* 50.93 mH, 0.6 lag, 161.24 V)

6. A coil when connected to 250 V d.c. supply dissipates 3125 watt of power. When connected across 250 V a.c. supply of frequency 50 c.p.s., it dissipates 1250 watt of power. Calculate the value of resistance and inductance of the coil. (*Ans.* 20 Ω, 0.078 H)

Type-III: Test Questions

1. Show that current lags behind the voltage vector in an *RL* series circuit. Also show that in *RL* series circuit, power, $P = VI \cos \phi$ where V and I are the rms values of voltage and current respectively.
2. Draw the phasor and wave diagram for voltage and current in an *RL* series circuit.
3. What do you mean by active and reactive component of alternating current?
4. What is power factor? Give its significance.
5. Why it is suggested to the industries to install a power factor improvement equipment in their prremises.

7.25 R – C SERIES CIRCUIT

A circuit that contains a pure resistance R ohm connected in series with a pure capacitor of capacitance C farad is known as $R-C$ series circuit.

An $R-C$ series circuit and its phasor diagram is shown in Fig. 7.34 and 7.35 respectively. To draw, the phasor diagram, current I (*r.m.s.* value) is taken as the reference vector. Voltage drop in resistance V_R $(= IR)$ is taken in phase with current vector, whereas, voltage drop in capacitive reactance V_C $(= IX_C)$ is taken 90° behind the current vector (since current leads the voltage by 90° in pure capacitive circuit). The vector sum of these two voltage drops is equal to the applied voltage V (r.m.s. value).

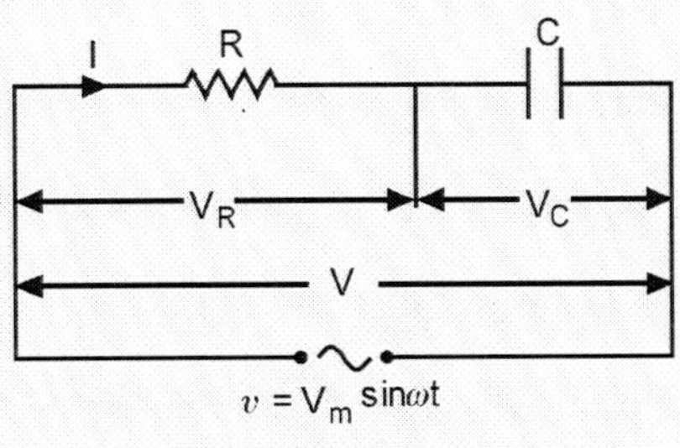

Fig. 7.34

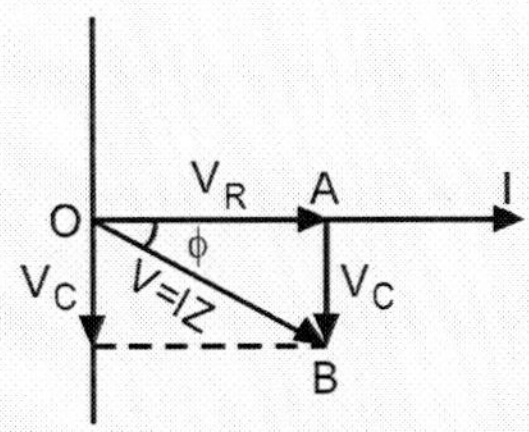

Fig. 7.35

Now $V_R = IR$ and $V_C = IX_C$ (where $X_C = 1/2\,\pi f C$)

In right angled triangle OAB

$$V = \sqrt{(V_R)^2 + (V_C)^2} = \sqrt{(IR)^2 + (IX_C)^2} = I\sqrt{R^2 + X_C^2}$$

or

$$I = \frac{V}{\sqrt{R^2 + X_C^2}} = \frac{V}{Z}$$

where $Z = \sqrt{R^2 + X_C^2}$ is the total opposition offered to the flow of alternating current by an $R - C$ series circuit and is called *impedance* of the circuit. It is measured in ohm.

Phase angle: From the phasor diagram it is clear that current in this circuit leads the applied voltage by an angle ϕ called *phase angle*.

From the phasor diagram shown in Fig. 7.35;

$$\tan\phi = \frac{V_C}{V_R} = \frac{IX_C}{IR} = \frac{X_C}{R} \quad \text{or} \quad \phi = \tan^{-1} X_C/R$$

Power: If the alternating voltage applied across the circuit is given by the equation:

$$v = V_m \sin \omega t \quad \text{...(i)}$$

Then,

$$i = I_m \sin(\omega t + \phi) \quad \text{...(ii)}$$

∴ Instantaneous power,

$$p = vi = V_m \sin \omega t \, I_m \sin(\omega t + \phi) = \frac{V_m I_m}{2} 2 \sin(\omega t + \phi) \sin \omega t$$

$$= \frac{V_m}{\sqrt{2}} \frac{I_m}{\sqrt{2}} [\cos\phi - \cos(2\omega t + \phi)] = \frac{V_m}{\sqrt{2}} \frac{I_m}{\sqrt{2}} \cos\phi + \frac{V_m}{\sqrt{2}} \frac{I_m}{\sqrt{2}} \cos(2\omega t + \phi)$$

Average power consumed in the circuit over a complete cycle,

$$P = \text{average of } \frac{V_m}{\sqrt{2}} \frac{I_m}{\sqrt{2}} \cos\phi - \text{average of } \frac{V_m}{\sqrt{2}} \frac{I_m}{\sqrt{2}} \cos(2\omega t + \phi)$$

or

$$P = \frac{V_m}{\sqrt{2}} \cdot \frac{I_m}{\sqrt{2}} \cos\phi - \text{zero} = V_{r.m.s.} I_{r.m.s.} \cos\phi = VI\cos\phi$$

where $\cos\phi$ is called *power factor* of the circuit.

From phasor diagram

$$\cos\phi = \frac{V_R}{V} = \frac{IR}{IZ} = \frac{R}{Z} \quad \text{same as in } R - L \text{ series circuit.}$$

Alternatively; Power,

$$P = VI\cos\phi = IZ \cdot I \cdot \frac{R}{Z} = I^2 R$$

This shows that power is actually consumed in resistance only; capacitor does not consume any power.

Power curve: The phasor diagram and wave diagram for voltage and current are shown in Fig. 7.36 (*a*) and 7.36 (*b*) respectively, where applied voltage ($v = V_m \sin wt$) is taken as reference quantity. The power curve for $R - C$ circuit is also shown in 24 (*b*). The points on the power curve are obtained from the product of the corresponding instantaneous values of voltage and current. It is clear that power is negative between angle (180° – f) and 180° and between (360° – f) and 360°. During rest of the cycle, the power is positive. Since the area under the positive loops is greater than that under the negative loops, the net power over a complete cycle is positive. Hence, a definite quantity of power is utilised or consumed by this circuit.

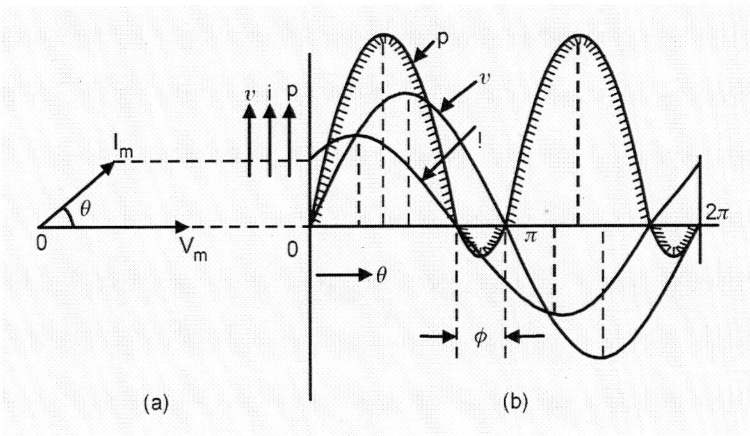

Fig. 7.36

Impedance triangle

When each side of the simplified phasor diagram shown in Fig. 7.37 is divided by a common factor *I*, we get another right angled triangle (shown in Fig. 7.38) known as *impedance triangle*.

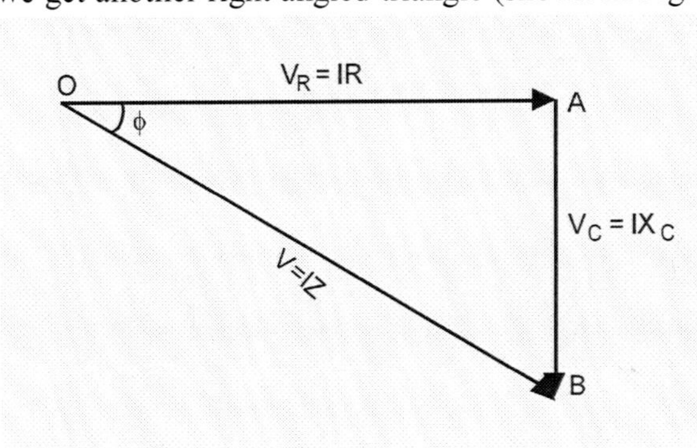

Fig. 7.37

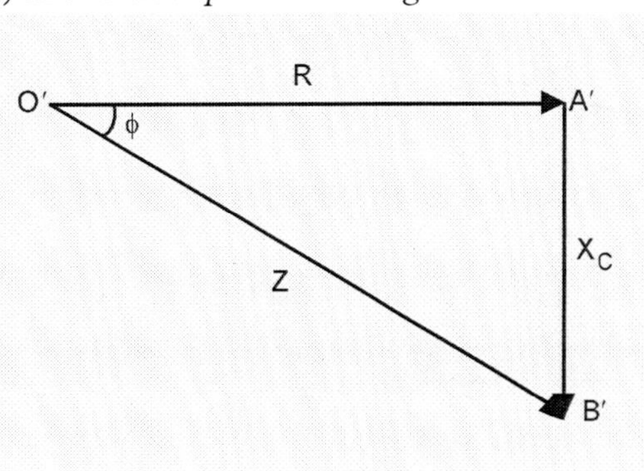

Fig. 7.38

Example 7.15. *A resistance of 15 ohm and capacitor of 150 μ F capacitance are connected in series across a 230 V, 50 Hz supply. Calculate (i) impedance of the circuit; (ii) current; (iii) power factor and phase angle; (iv) power consumed in the circuit.*

Solution: The circuit is shown in Fig. 7.39.

Impedance, $Z = \sqrt{R^2 + X_C^2}$

where, $X_C = 1/2\,\pi f C = 1/2\,\pi \times 50 \times 150 \times 10^{-6}$

$= 21.22\ \Omega;$

$R = 15\ \Omega$

(*i*) $\therefore\ Z = \sqrt{(15)^2 + (21\cdot22)^2}$ = **25.987 Ω** (*Ans.*)

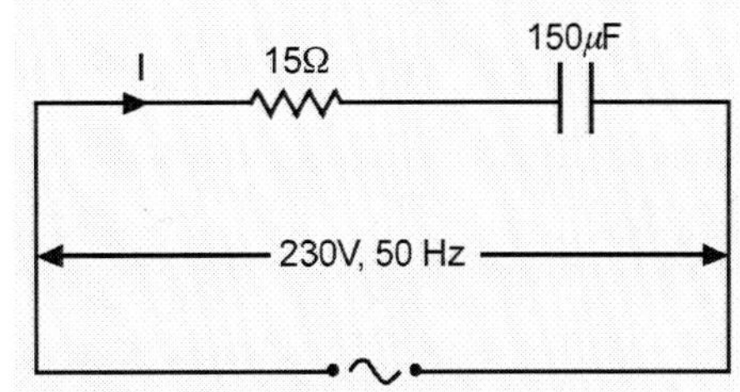

Fig. 7.39

(*ii*) Current, $I = \frac{V}{Z} = \frac{230}{25 \cdot 987} = \mathbf{8.85\ A}$ (*Ans.*)

(*iii*) Power factor, $\cos\phi = \frac{R}{Z} = \frac{15}{25 \cdot 987} = \mathbf{0.577\ leading}$ (*Ans.*)

Phase angle, $\phi = \cos^{-1} 0.577 = \mathbf{54.75°}$ (*Ans.*)

(*iv*) Power, $P = VI\cos\phi = 230 \times 8.85 \times 0.577 = \mathbf{1174.9\ W}$ (*Ans.*)

Example 7.16. *A voltage of 125 V at 50 Hz is applied across a non inductive resistor connected in series with a condenser. The current in the circuit is 2.2 A. The power loss in the resistor is 96.8 W and that in the condenser is negligible. Calculate the resistance and the capacitance.*

Solution: Applied voltage, $V = 125$ V; Supply frequency, $f = 50$ Hz
Current in the circuit, $I = 2.2$ A; Power loss in the resistor, $P = 96.8$ W
Power loss in resistor, $P = I^2R$

$\therefore$ Circuit resistance, $R = \frac{P}{I^2} = \frac{96 \cdot 8}{2 \cdot 2 \times 2 \cdot 2} = \mathbf{20\ ohm}$ (*Ans.*)

Circuit impedance, $Z = \frac{V}{I} = \frac{125}{2 \cdot 2} = 56.82$ ohm

Capacitive reactance, $X_C = \sqrt{Z^2 - R^2} = \sqrt{(56 \cdot 82)^2 - (20)^2} = 53.18$ ohm

Capacitance of capacitor, $C = \frac{1}{2\pi f X_C} = \frac{1}{2\pi \times 50 \times 53 \cdot 18} = \mathbf{59.85\ \mu F}$ (*Ans.*)

Example 7.17. *A 120 V, 60 W lamp is to be operated on 220 V, 50 Hz supply mains. In order that lamp should operate in correct voltage. Calculate value of:*
(*i*) *non inductive resistance* (*ii*) *pure inductance*

Ans. Lamp's rating: 120 V, 60 W

Supply voltage, $V_S = 220$ V and Frequency, $f = 50$ Hz

Rsistance of the lamp, $R_L = \frac{(120)^2}{60} = 240\ \Omega$

Operating current, $I = \frac{60}{120} = 0 \cdot 5$ A

(*i*) For operating the lamp using non-inductive resistance, as shown in Fig. 7.40 (*a*). Let the value of resistance be *R* W.

$\therefore \quad I(R + R_L) = V$

or $\quad R + R_L = \frac{220}{0 \cdot 5} = 440$

or $\quad R = 440 - R_L = 440 - 240 = \mathbf{200\ \Omega}$ (*Ans.*)

(*ii*) For opesrating the lamp using pure inductance as shown in Fig. 7.40 (*b*). Let the value of inductance be *L* heavy and $X_L = 2\,p\,fL$,

$$\therefore \quad IZ = V \quad \text{or} \quad Z = \frac{V}{I} = \frac{220}{0\cdot 5} = 440\ \Omega$$

$$\text{or} \quad \sqrt{R_L^2 + X_L^2} = 440$$

$$R_L^2 + X_L^2 = (440)^2 \quad \text{or} \quad X_L^2 = 440^2 - 240^2$$

$$\text{or} \quad X_L = \sqrt{440^2 - 240^2} = 368.78\ \Omega$$

$$\text{or} \quad 2\pi f L = 368.78 \quad \text{or} \quad L = \frac{368\cdot 78}{2\pi \times 50} = \mathbf{1.174\ H}\ (Ans)$$

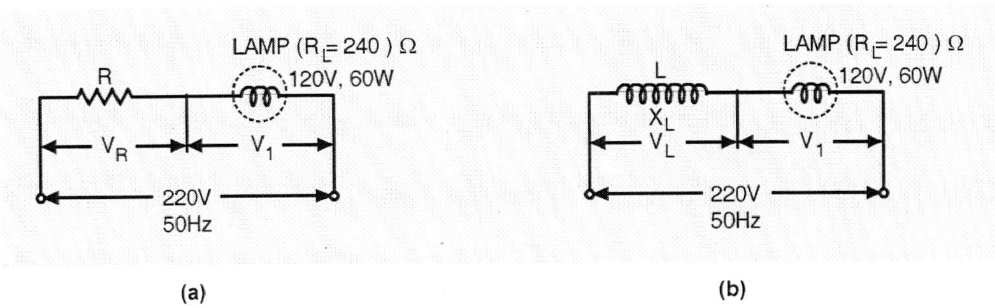

Fig. 7.40

Example 7.18. *A 400 Hz generator has an induced e.m.f. of 100 V and an internal impedance (5 + j0) ohm. If it supplies an impedance consisting of a 40 ohm capacitive reactance in series with 10 ohm resistance, what is the magnitude of the current passing through it? Determine also the voltage at the terminals of the generator.*

Solution: Simple circuit is shown in Fig. 7.41.

Total impedance of the circuit.

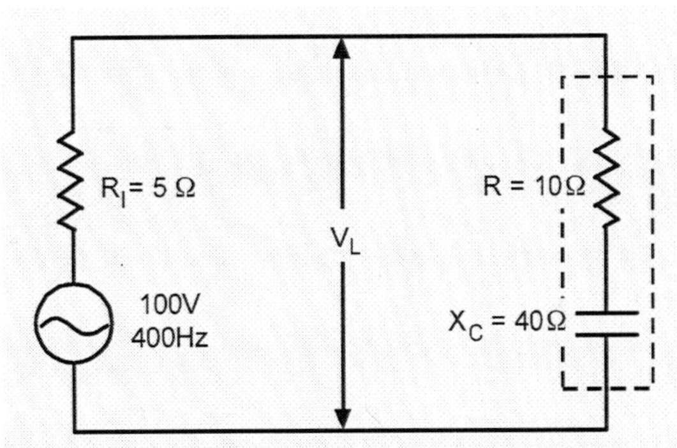

Fig. 7.41

$$Z_T = \sqrt{(R_i + R)^2 + (X_C)^2}$$

$$= \sqrt{(5 + 10)^2 + (40)^2} = 42.72\ \Omega$$

Circuit current, $I = V/Z = 100/42.72 = \mathbf{2.341\ A}$ (*Ans.*)

Load impedance, $Z_L = \sqrt{R^2 + X_C^2} = \sqrt{(10)^2 + (40)^2} = 41.231\ \Omega$

Terminal voltage, $V_L = IZ_L = 2.341 \times 41.231 = \mathbf{96.52\ V}$ (*Ans*).

7.26 R – L – C SERIES CIRCUIT

A circuit that contains a pure resistance of *R* ohm, a pure inductance of *L* henry and a pure capacitor of capacitance *C* farad; all connected in series is known as *R – L – C* series circuit.

An $R-L-C$ series circuit is shown in Fig. 7.42.

Here $X_L = 2\,\pi f L$ and $X_C = 1/2\,\pi f C$

When a resulting current I (r.m.s. value) flows through the circuit, the voltage across each component will be

$V_R = IR$ *i.e.* voltage across R........... in phase with I;

$V_L = IX_L$ *i.e.* voltage across L........... leads I by 90°;

$V_C = IX_C$ *i.e.* voltage across C........... lags I by 90°;

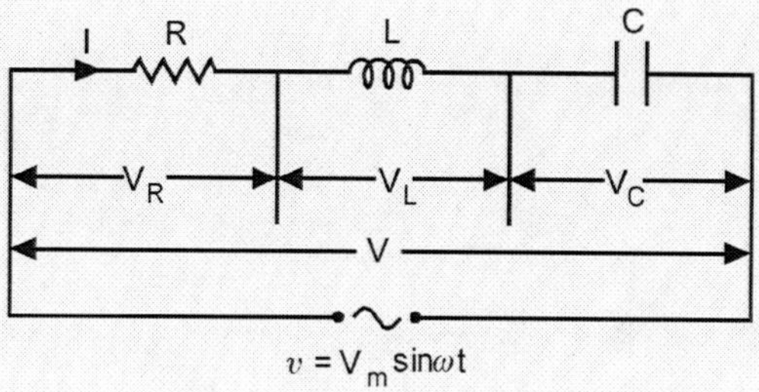

Fig. 7.42

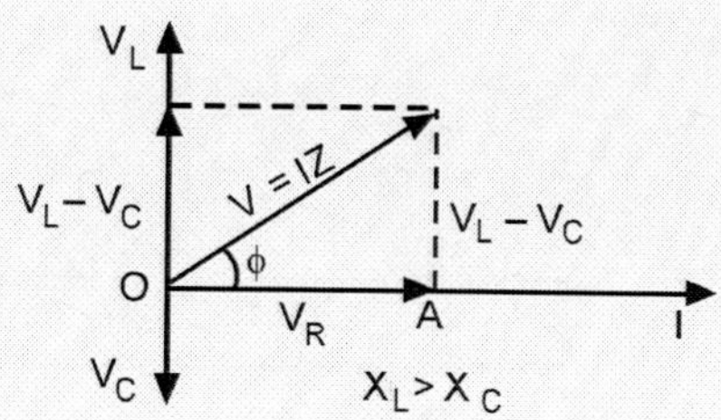

Fig. 7.43

The phasor diagram is shown in Fig. 7.43 where current is taken as the reference phasor. Since voltage across inductance V_L leads the current vector I by 90° and voltage across capacitance V_C lags the current vector I by 90°, they act opposite to each other. If $V_L > V_C$, in effect, the circuit behaves as an inductive circuit but when $V_L < V_C$, the circuit behaves as a capacitive circuit. Here, the phasor diagram is drawn for an inductive circuit (*i.e.* when $V_L > V_C$).

$$V = \sqrt{(V_R)^2 + (V_L - V_C)^2} = \sqrt{(IR)^2 + (IX_L - IX_C)^2}$$

or $$V = I\sqrt{R^2 + (X_L - X_C)^2}$$

or $$I = \frac{V}{\sqrt{(R)^2 + (X_L - X_C)^2}} = \frac{V}{Z}$$

where $Z = \sqrt{R^2 + (X_L - X_C)^2}$ is the total opposition offered to the flow of alternating current by an $R-L-C$ series circuit and is called *impedance* of the circuit.

Phase angle: From phasor diagram: $\tan\phi = \dfrac{V_L - V_C}{V_R} = \dfrac{X_L - X_C}{R}$

or $$\phi = \tan^{-1}\frac{X_L - X_C}{R}$$

Power: Average power, $P = VI\cos\phi = I^2 R$

Power, factor, $\cos\phi = \dfrac{V_R}{V} = \dfrac{R}{Z}$

If the alternating voltage applied across the circuit is given by the equation

$$v = V_m \sin \omega t$$

the circuit current is represented by the equation as per the constants or parameters explained below:

Three cases of $R - L - C$ series circuit

(*i*) When $X_L > X_C$, the phase angle *f* is positive. In effect, the circuit behaves as an $R - L$ series circuit. The circuit current lags behind the applied voltage and p.f. is lagging. The current is given by the equation.

$$i = I_m \sin (\omega t - \phi)$$

(*ii*) When $X_L < X_C$, the phase angle *f* is negative. In effect, the circuit behaves as an $R - C$ series circuit. The circuit current leads the applied voltage and p.f. is leading. The current is given by the equation.

$$i = I_m \sin (\omega t + \phi)$$

(*iii*) When $X_L = X_C$, the phase angle *f* is zero. In effect, the circuit behaves like a pure resistive circuit. The circuit current is in phase with applied voltage and p.f. is unity. The current is given by the equation.

$$i = I_m \sin \omega t$$

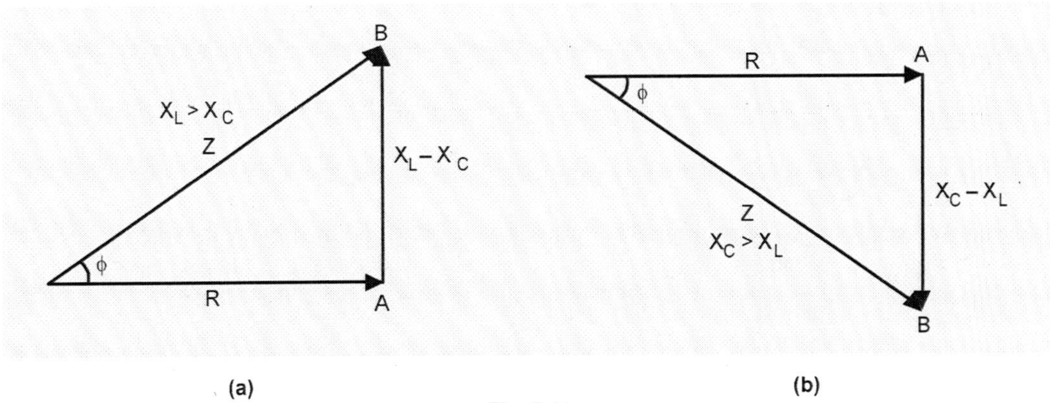

Fig. 7.44

Impedance triangle

Fig. 7.44 (*a*) shows the impedance triangle of the circuit when $X_L > X_C$, whereas, Fig. 7.44 (*b*) shows the impedance triangle of the circuit when $X_L < X_C$.

7.27. SERIES RESONANCE

In an $R - L - C$ series circuit, when circuit current is in phase with the applied voltage, the circuit is said to be in series resonance. This condition is obtained in an $R - L - C$ circuit, shown in Fig. 7.45,

when $X_L = X_C$ (or $X_L - X_C = 0$)

At resonance; $X_L - X_C = 0$ or $X_L = X_C$

Impedance, $Z_r = \sqrt{R^2 + (X_L - X_C)^2} = R$

Current, $I_r = \dfrac{V}{Z_r} = \dfrac{V}{R}$

Since, at resonance, the opposition to the flow of current is only resistance (R) of the circuit, the circuit draws *maximum current* under this conditions.

Resonant frequency: The value of X_L (= $2\pi f L$) and X_C (= $1/2\pi f C$) can be changed by changing the supply frequency. When frequency increases, the value of X_L, increases, whereas, the value of X_C decreases, and *vice-versa*. Thus to obtain series resonance, the frequency is adjusted to f_r so that $X_L = X_C$, the condition at point P shown in Fig. 7.46.

∴ At series resonance, $X_L = X_C$

$$2\phi f_r L = \frac{1}{2\pi f_r C} \quad \text{or} \quad f_r = \frac{1}{2\pi\sqrt{LC}}$$

where f_r is the *resonant frequency* in Hz when L and C are measured in henry and farad respectively.

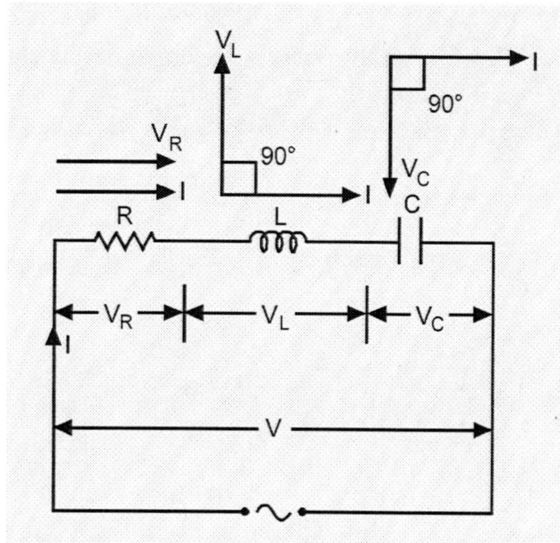

Fig. 7.45.

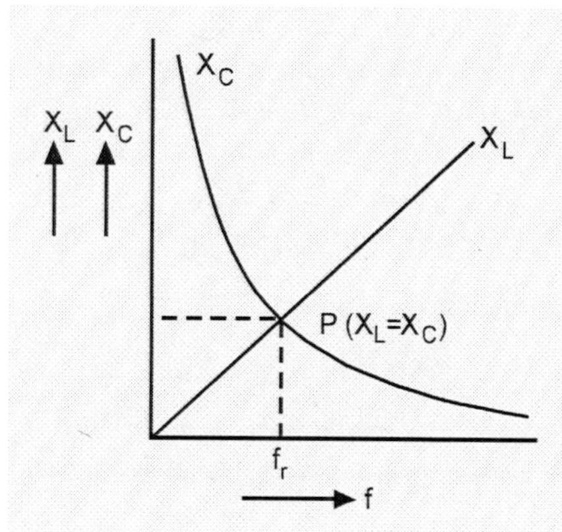

Fig. 7.46

Effects of series resonance

The following are the main effects of series resonance:

(*i*) At resonance $X_L = X_C$, therefore, the impedance of the circuit is minimum and is reduced to the resistance of the circuit only, *i.e.*

$$Z_r = R$$

(*ii*) Since, impedance is minimum, the circuit current is maximum at resonance, *i.e.*

$$I_r = V/Z_r = V/R$$

(*iii*) Power taken by the circuit is maximum, as I_r is maximum, *i.e.*

$$P_r = I_r^2 R$$

(*iv*) As the current drawn by the circuit, at resonance, is very large (maximum), the voltage drop across L (*i.e.* $V_L = IX_L = I \times 2pf_r L$) and C (*i.e.* $V_C = IX_C = I \times 1/2pf_r C$) are also very large.

In power system, at resonance, the excessive voltage built up across the inductive and capacitive components (such as circuit breakers, reactors, etc.) may cause damage. Therefore, series resonance should be avoided in power system. However, in some of the electronic devices (such as antenna circuit of radio and *TV* receiver, tuning circuits etc.), the principle of series resonance is used to increase the signal voltage and current at a desired frequency (f_r).

Since a series resonant circuit has the capability to draw heavy current and power from the mains, it is often regarded as **acceptor circuit.**

7.28 Q-FACTOR OF SERIES RESONANT CIRCUIT

We have seen that at series resonance, the circuit draws largest current from the mains, this produces a heavy voltage across L or C. The factor by which the *p.d.* across L or C rises to that of the applied voltage is called the *Q-factor* of the series resonant circuit.

$$\therefore \quad Q\text{-factor} = \frac{\text{Voltage across } L \text{ or } C}{\text{Applied voltage}} = \frac{I_r X_L}{I_r R} = \frac{X_L}{R} = \frac{\omega_r L}{R}$$

$$\text{where, } \omega_r = 2\pi f_r = 2\pi \frac{1}{2\pi\sqrt{LC}} = \frac{1}{\sqrt{LC}}$$

$$\therefore \quad Q\text{-factor} = \frac{L}{R} \times \frac{1}{\sqrt{LC}} = \frac{1}{R}\sqrt{\frac{L}{C}}$$

The value of Q-factor depends entirely upon the design of coil (*i.e.* $R - L$ which is a part of $R - L - C$ circuit).

Example 7.19. *A coil resistance* 10W and inductance 0.14 *H is connected in series with a capacitor of* 150 *mF across a* 200 *V*, 50 *Hz supply. Calculate* (*a*) *inductive reactance* (*b*) *capacitive reactance* (*c*) *impedance* (*d*) *current* (*e*) *voltage across coil and capacitor.*

Solution: The circuit is shown in Fig. 7.47.

Where, $R = 10\ \Omega$; $L = 0.14$ H; $C = 150\ \mu\text{F} = 150 \times 10^{-6}$ F; $V = 200$ V; $f = 50$ Hz

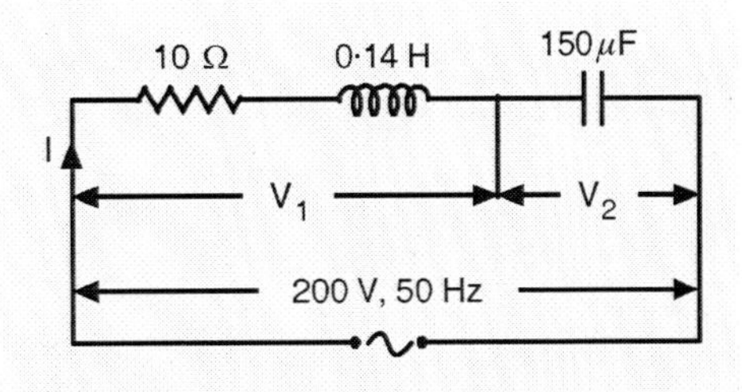

Fig. 7.47

(*a*) Inductive reactance, $X_L = 2\pi f L = 2\pi \times 50 \times 0.14 = 44\ \Omega$ (*Ans.*)

(*b*) Capacitive reactance, $X_C = \dfrac{1}{2\pi f C}$

$$= \frac{1}{2\pi \times 50 \times 150 \times 10^{-6}} = 21.22\ \Omega \text{ (Ans.)}$$

(*c*) Impedance, $Z = \sqrt{R^2 + (X_L - X_C)^2} = \sqrt{10^2 + (44 - 21\cdot22)^2}$

$= 24.88\ \Omega$ (*Ans.*)

(*d*) Current, $I = \dfrac{V}{Z} = \dfrac{200}{24\cdot88} = 8.04$ A (*Ans.*)

(*e*) Voltage across coil, $V_1 = I Z_{\text{COIL}} = I\sqrt{R^2 + X_L^2}$

$= 8\cdot04\sqrt{10^2 + 44^2} = 362.8$ V (*Ans.*)

Voltage across capacitor, $V_2 = I X_C = 8.04 \times 21.22 = 170.6$ V (*Ans.*)

Example 7.20. *A coil of resistance* 8W *and inductance* 0.12 *henry is connected in series with a condenser of capacitance* 140 *microfarad across a 230 volt, 50 Hz supply. Determine:*

(*i*) *Impedance of the entire circuit,*

(*ii*) *Current flowing through the condenser,*

(*iii*) *Power factor of the circuit,*

(*iv*) *Voltage across the condenser.*

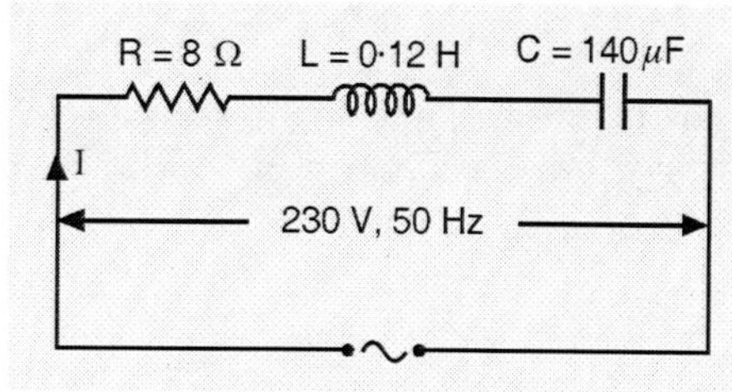

Fig. 7.48.

Solution: The circuit is shown in Fig. 7.48.

where, Resistance of the coil, $R = 8\ \Omega$

Inductive reactance of the coil, $X_L = 2\pi f L$

$= 2\pi \times 50 \times 0.12 = 37.7\ \Omega$

$$\text{Capacitive reactance of the condenser, } X_C = \frac{1}{2\pi f C} = \frac{1}{2\pi \times 50 \times 140 \times 10^{-6}} = 22.74\ \Omega$$

(*i*) Impedance of the whole circuit,

$$Z = \sqrt{R^2 + (X_L - X_C)^2} = \sqrt{8^2 + (37\cdot7 - 22\cdot74)^2}$$

$$= \mathbf{17\ \Omega}\ (Ans.)$$

(*ii*) Current flowing through the condenser,

$$I_C = \text{Circuit current, } I = \frac{V}{Z} = \frac{230}{17} = \mathbf{13.53\ A}\ (Ans.)$$

(*iii*) Power factor of the circuit, $\cos\phi = \frac{R}{Z} = \frac{8}{17} = \mathbf{0.471\ (lagging)}\ (Ans.)$

$\because X_L > X_C$

(*iv*) Voltage across condenser, $V_C = I_C X_C = 13.53 \times 22.74 = \mathbf{307.7\ V}\ (Ans.)$

Example 7.21. *Find applied voltage and power loss in the circuit shown in Fig. 7.49.*

Solution: Here, $C = 20\ \mu F$; $I = 0.345$ A

Applied voltage, $V = \sqrt{25^2 + (40 - 55)^2} = \mathbf{29.155\ V}\ (Ans.)$

Power Loss, $P = V_R \times I = 25 \times 0.345 = \mathbf{8.625\ watt}\ (Ans.)$

Example 7.22. *A coil of resistance* 12 Ω *and inductance* 0.05 *H, a non-inductive resistance of* 20 Ω *resistance and a loss-free* 40 μF *capacitor are connected across a* 240 *V,* 50 *Hz sinusoidal supply. Calculate* (*i*) *the current and* (*ii*) *the power factor of the circuit.*

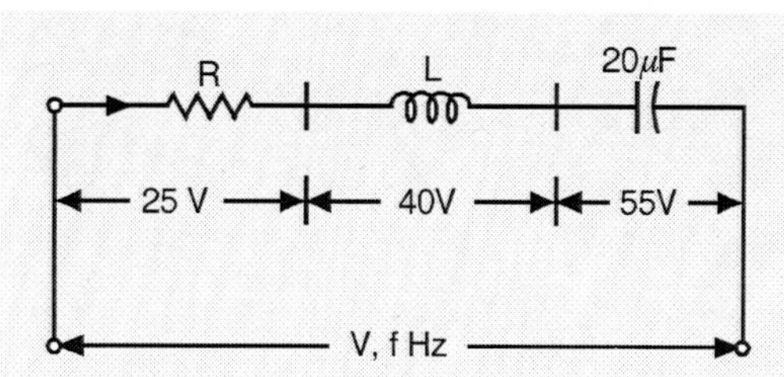

Fig. 7.49

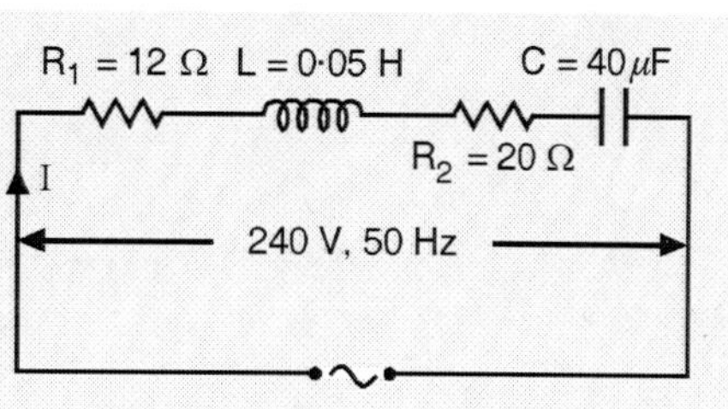

Fig. 7.50

Solution: The circuit is shown in Fig. 7.50.

Here, $R_1 = 12\ \Omega$; $L = 0.05$ H; $R_2 = 20\ \Omega$;

$C = 40\ \mu F = 40 \times 10^{-6}$ F; $V = 240$ V; $f = 50$ Hz

Resistance of the whole circuit, $R = R_1 + R_2$

$= 12 + 20 = 32\ \Omega$

Inductive reactance, $X_L = 2\pi f L$

$= 2\pi \times 50 \times 0.05$

$= 15.7\ \Omega$

$$\text{Capacitive reactance, } X_C = \frac{1}{2\pi f C}$$

$$= \frac{1}{2\pi \times 50 \times 40 \times 10^{-6}} = 79 \cdot 58\ \Omega$$

$$\text{Impedance, } Z = \sqrt{R^2 + (X_L - X_C)^2}$$

$$= \sqrt{32^2 + (15 \cdot 7 - 79 \cdot 58)^2} = 71 \cdot 45\ \Omega$$

(*i*) Circuit current, $I = \frac{V}{Z} = \frac{240}{71 \cdot 45}$ = **3.36 A** (*Ans.*)

(*ii*) Power factor of the circuit, $\cos\phi = \frac{R}{Z} = \frac{32}{71 \cdot 45}$ = **0.448 (leading)** (*Ans.*)

(since $X_C > X_L$)

Example 7.23. *A series R-L-C circuit consisting of a resistance of* 20 Ω, *inductance* 0.2 H *and capacitance of* 150 *μ F is connected across a* 230 *V*, 50 *Hz source. Calculate* (*i*) *the impedance* (*ii*) *the current* (*iii*) *the magnitude and nature of the power factor* (*iv*) *the frequency of supply to be adjusted to make power factor unity.*

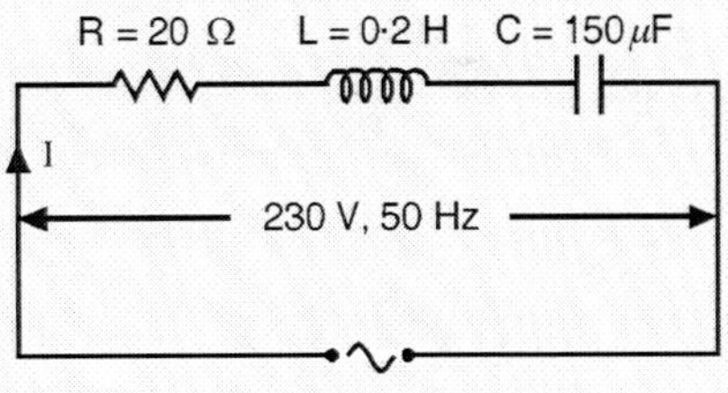

Fig. 7.51

Solution: The circuit is shown in Fig. 7.51.

Here, $R = 20\ \Omega$; $L = 0.2$ H; $C = 150\ \mu F = 150 \times 10^{-6}$ F

Inductive, reactance, $X_L = 2\pi f L = 2\pi \times 50 \times 0.2 = 62.83\ \Omega$

$$\text{Capacitance reactance, } X_C = \frac{1}{2\pi f C} = \frac{1}{2\pi \times 50 \times 150 \times 10^{-6}} = 21.22\ \Omega$$

$$(i) \text{ Impedance, } Z = \sqrt{R^2 + (X_L - X_C)^2}$$

$= \sqrt{20^2 + (62 \cdot 83 - 21 \cdot 22)^2}$ = **46.17 Ω** (*Ans.*)

(*ii*) Circuit current, $I = \frac{V}{Z} = \frac{230}{46 \cdot 17}$ = **4.98 A** (*Ans.*)

$$(iii)\ \text{Power factor} = \cos\phi = \frac{R}{Z} = \frac{20}{46\cdot 167} = \mathbf{0.433}\ (Ans.)$$

The power factor, is **lagging** because inductive reactance is more than capacitive reactance *i.e.* the circuit behaves as an inductive circuit.

Power factor will be unity when

$$X_L = X_C \quad \text{or} \quad 2\pi f_r L = \frac{1}{2\pi f_r C}$$

or

$$f_r = \frac{1}{2\pi}\frac{1}{\sqrt{LC}} = \frac{1}{2\pi\sqrt{0\cdot 2 \times 150 \times 10^{-6}}} = \mathbf{29.06\ Hz}\ (Ans.)$$

Example 7.24. *An inductive coil takes 10 A and dissipates 1000 W when connected to a 250 V, 25 Hz supply. Calculate the following, (i) the impedance; (ii) the effective resistance; (iii) the reactance; (iv) the value of the capacitance required to be connected in series with coil to make the power factor of the circuit unity; and (vi) what is now the current taken by the coil? Also draw the phasor diagram of the two cases.*

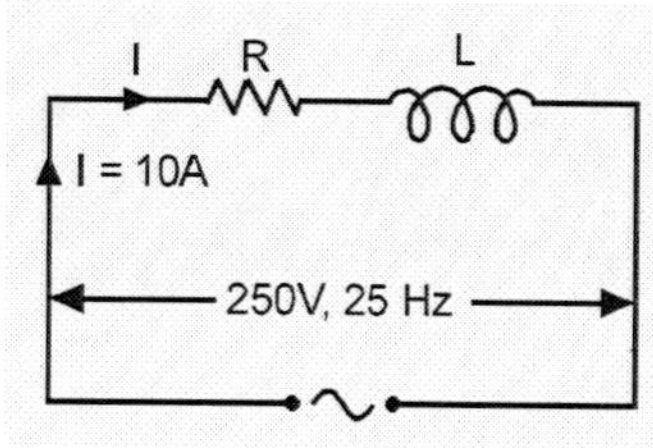

Fig. 7.52

Solution: The circuit is shown in Fig. 7.52.

Power dissipated in the coil, $I^2 R = 1000$ W

$\therefore$ Resistance of the coil, $R = 1000/(10)^2 = 10\ \Omega$ (*Ans.*)

Impedance of the coil, $Z = V/I = 250/10 = 25\ \Omega$ (*Ans.*)

Inductive reactance of the coil, $X_L = \sqrt{Z^2 - R^2} = \sqrt{(25)^2 - (10)^2} = \mathbf{22.91\ \Omega}$ (*Ans.*)

Power factor, $\cos\phi = R/Z = 10/25 = \mathbf{0.4\ lag}$ (*Ans.*)

The p.f. of the circuit will be unity when, $X_C = X_L$

or $\quad 1/2\pi f C = 22.91$ or $C = \dfrac{1}{2\pi \times 25 \times 22\cdot 91} = \mathbf{277.84\ \mu F}$ (*Ans.*)

Now, $\quad$ current, $I = \dfrac{V}{R} = \dfrac{250}{10} = \mathbf{25\ A}$ (*Ans.*)

The phasor diagram for the two cases is drawn in Fig. 7.53 (*a*) and (*b*) respectively.

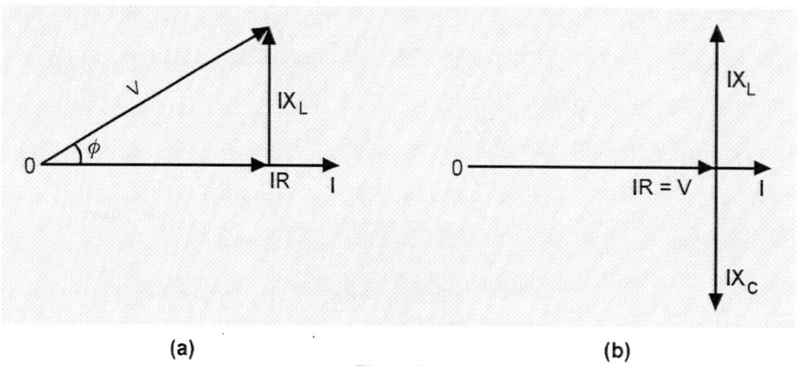

Fig. 7.53

Example 7.25. *A choke coil is connected series with a 200 μF capacitor. With a constant supply voltage of 250 V, it is found that the circuit takes its maximum current of 50 A when the supply frequency is 100 Hz. Determine (i) resistance and inductance of the choke coil; (ii) voltage across the capacitor and (iii) Q-factor of the circuit.*

Solution: At resonance; Current, $I_r = V/R$

or $$R = V/I_r = 250/50 = 5\ \Omega \text{ (Ans.)}$$

Also $$2\pi f_r L = 1/2\pi f_r C$$

or $$L = \frac{1}{(2\pi f_r)^2 C} = \frac{1}{(2\pi \times 100)^2 \times 200 \times 10^{-6}} = \mathbf{12.66\ mH} \text{ (Ans.)}$$

Voltage across capacitor,

$$V_C = I_r X_C = 50 \times 1/2\pi \times 100 \times 200 \times 10^{-6} = \mathbf{397.88\ V} \text{ (Ans.)}$$

Example 7.26. *A series R-L-C circuit with R = 10Ω; L = 0.02 H, C = 2 μF is connected to 100 V variable frequency source. Find the frequency for which the current is maximum.*

Solution: In *RLC* series circuit, the current will be maximum when impedance is minimum *i.e.*

when $$Z = R \quad \text{or} \quad X_L = X_C \text{ (at resonance)}$$

or $$2\pi f_r L = \frac{1}{2\pi f_r C}$$

or $$f_r = \frac{1}{2\pi\sqrt{LC}} = \frac{1}{2\pi\sqrt{0\cdot 02 \times 2 \times 10^{-6}}} = \mathbf{795.77\ Hz} \text{ (Ans.)}$$

TEST YOUR LEARNING-3

Type-I: Short Answer Questions

1. What do you mean by impedance of an ac circuit?
2. Draw an impedance triangle for *RL*, *RC* and *RLC* series circuit and label it.
3. Define power factor of an ac circuit.
4. In an *RLC* series circuit, show that power is consumed only by resistor.
5. What do you mean by series resonance?
7. Define Q-factor of a series resonant circuit.
8. When does an *RLC* series circuit called as inductive circuit or capacitive circuit?

Type-II: Numerical for Practice

1. A coil of power factor 0.6 is in series with a 100 μ F capacitor, when connected to 50 Hz supply the potential difference across the coil is equal to the potential difference across the capacitor. Find the resistance and inductance of the coil. (**Ans.** *19.098 Ω; 81.06 mH*)
2. A voltage of 125 V at 50 Hz is applied across a non-inductive resistor connected in series with a condenser. The current in the circuit is 2.2 A. The power loss in the resistor is 96.8 W and that is the condenser is negligible. Calculate the resistance and the capacitance. (**Ans.** *20 Ω; 59.85 μ F*)

3. A metal filament lamp, rated at 750 W, 100 V is to be connected in series with a capacitor across a 230 V, 60 Hz supply. Calculate: (*i*) The capacitance required; (*ii*) The power factor.

(**Ans.** *96.08 μ F; 0.4346 leading*)

4. An a.c. circuit having a resistance of 10 ohm, an inductance of 0.2 henry and a capacitance of 100 μF in series, is connected across a single-phase 110V, 50 Hz supply. Calculate the following for this circuit. (*i*) Resultant reactance; (*ii*) Impedance; (*iii*) Current; (*iv*) Voltages across *R*, *L* and *C*; (*v*) Phase difference between the current and supply voltage; (*vi*) Draw the phasor diagram of the circuit.

(**Ans.** *31 ohm; 32.573 ohm; 3.377 A, 33.77V, 212.18 V, 107.49 V; 72.12°*)

5. A coil of p.f. 0.6 is in series with a 80μ F capacitor. When connected to a 60 Hz supply, the voltage across the capacitor is equal to the voltage across the coil. Find the resistance and inductance of the coil. (**Ans.** *19.89 Ω; 70.36 m H*)

Type-III: Test Questions

1. Show that current vector leads the voltage vector in an *RC* series circuit. Also show that in an *RC* series circuit, power $P = VI \cos \phi$ where *V and I* are the *rms* values of voltage and current respectively.
2. Draw the phasor and wave diagram for voltage and current in an *RC* series circuit.
3. What is power factor in an *RC* series circuit and explain how it is different to *RL* series circuit

SUMMARY

1. *Alternating voltage*: A voltage that changes its polarity and magnitude at regular intervals of time is called an *alternating voltage*.
2. *Sinusoidal alternating quantity*: An alternating quantity that varies according to sine of angle θ is known as sinusoidal alternating quantity.
3. *Equation of an alternating quantity*:

$$e = E_m \sin \theta = E_m \sin \omega t; \; i = I_m \sin \theta = I_m \sin \omega t$$

4. *Wave form*: The shape of the curve obtained by plotting the instantaneous values along *y*-axis and time or angle $\theta (= \omega t)$ along *x*-axis is called wave form or wave shape.
5. *Cycle*: When an alternating quantity goes through a complete set of +ve and –ve value or goes through 360 electrical degrees, it is said to have completed a cycle.
6. *Alternation*: One half of cycle is called an alternation.
7. *Time period* (*T*): Time taken to complete one cycle in second is called time period.
8. *Frequency* (*f*): The number of cycles made per second by an alternating quantity is called its frequency. It is measured in c/s or Hz (Hertz).
9. *Relations*: $f = 1/T$ and $\omega = 2 \pi f$
10. Instantaneous value (υ or i): The value of an alternating quantity at any instant is called its instantaneous value.

11. *Amplitude* (V_m or I_m): The maximum value obtained by an alternating quantity during a cycle is called its amplitude or maximum value or peak value or crest value.

12. *Average value* (V_{av} or I_{av}): The arithmetic average of all the instantaneous values considered of an alternating quantity over one cycle is called its *average value*.

$$I_{av} = \frac{i_1 + i_2 + i_3 + ... + i_n}{n}$$

For sinusoidal current, $I_{av} = 2\, I_m/\pi = 0.637\, I_m$.

13. *Effective or r.m.s. value* ($V_{r.m.s.}$ or $I_{r.m.s.}$): That steady current which when flows through a resistor of known resistance for a given time produces the same amount of heat as produced by the alternating current when flows through the same resistor for the same time is called effective or r.m.s. value of the alternating current.

$$I_{r.m.s.} = \sqrt{\frac{i_1^2 + i_2^2 + i_3^2 + ... + i_n^2}{n}}$$

For sinusoidal current, $I_{r.m.s.} = I_m/\sqrt{2} = 0.707\, I_m$.

14. *Form factor:* $I_{r.m.s.}/I_{av}$; For sinusoidal quantities, its value is 1.11.

15. *Peak factor*: $I_m/I_{r.m.s.}$; For sinusoidal quantities, its value is 1.414.

16. *Phase*: The phase of an alternating quantity at an instant is defined as the fractional part of a cycle through which the quantity has advanced from a selected origin. It has less importance, in practice.

17. *Phase difference*: The angular displacement between the maximum positive values of two alternating quantities having the same frequency is called the phase difference between them.

Mathematically ; If $i_1 = I_{m1} \sin \omega t$ the $i_2 = I_{m2} \sin (\omega t \pm \phi)$

18. *Leading quantity*: An alternating quantity that attains its + ve maximum value prior to the other is called leading quantity.

19. *Lagging quantity*: An alternating quantity that attains its +ve maximum value after the other is called lagging quantity.

20. *A.C. Circuit:* The path for the flow of alternating current is called a.c. circuit.

21. *Pure resistive circuit*: Circuit contains pure resistance.
$v = V_m \sin \omega t$; $i = I_m \sin \omega t$; $I_m = V_m/R$; $P = VI$; $p.f. = 1$;
Current is in phase with voltage vector.

22. *Pure inductive circuit*: Circuit contains pure inductance.
$v = V_m \sin \omega t$; $i = I_m \sin (\omega t - \pi/2)$; $I_m = V_m/X_L$; $X_L = 2\, \pi f L$; $P = 0$; $p.f. = 0$ lag;
Current lags behind voltage vector of 90° (*i.e.* $\pi/2$)

23. *Pure capacitive circuit*: Circuit contains pure capacitance.
$v = V_m \sin \omega t$; $i = I_m \sin (\omega t + \pi/2)$; $I_m = V_m/X_C$; $X_C = 1/2\, \pi f C$; $P = 0$; $p.f. = 0$ lead;
Current leads the voltage vector by 90° (*i.e.* $\pi/2$).

24. *R—L series circuit*: Circuit contains resistance and inductance in series. It is generally called inductive circuit.

$v = V_m \sin \omega t$; $i = I_m \sin (\omega t - \phi)$; $\phi = \tan^{-1} X_L/R$; $I_m = V_m/Z$; $Z = \sqrt{R^2 + X_L^2}$;
$P = VI \cos \phi = I^2 R$; $p.f. = \cos \phi = R/Z$ lag;

25. *Power factor and its importance*: p.f. = $\cos \phi = R/Z$ = true power/apparent power
Ill-effects of poor p.f. (*i*) greater conductor size, (*ii*) poor efficiency, (*iii*) poor voltage regulation and (*iv*) larger *kVA* rating of equipment.

26. *R—C series circuit*: Circuit contains resistance and capacitance in series. It is generally called capacitive circuit.

$v = V_m \sin \omega t$; $i = I_m \sin (\omega t + \phi)$; $\phi = \tan^{-1} X_C/R$; $I_m = V_m/Z$; $Z = \sqrt{R^2 + X_C^2}$;
$P = VI \cos \phi = I^2 R$; P.f. = $\cos \phi = R/Z$ leading;

27. *R—L—C series circuit*: Circuit contains resistance, inductance and capacitance in series. This circuit behaves as under:
(*i*) When $X_L > X_C$..........inductive circuit..........p.f. lagging....0 to 1
(*ii*) When $X_L < X_C$..........capacitive circuit........p.f. leading....0 to 1
(*iii*) When $X_L = X_C$..........resistive circuit.........p.f. unity....1 only.

28. *Series resonance*: An $R - L - C$ circuit is said to be in series resonance, when

$X_L = X_C$; $Z_r = \sqrt{R^2 + (X_L - X_C)^2} = R$
Resonance frequency, $f_r = 1/2\pi\sqrt{LC}$
Current is maximum *i.e.* $I_r = V/Z_r = V/R$

$$\text{Q-factor} = 2\pi f_r L/R = \frac{1}{R}\sqrt{\frac{L}{C}}$$

SELF ASSESMENT
TEST YOUR PREPARATION

OBJECTIVE TYPE QUESTIONS

Type-I: Fill in the blanks:

1. Peak factor is defined as the ratio of __________ and __________.
2. In an a.c. circuit, an ammeter measures __________ value of current.
3. The number of cycles made per second by an alternating quantity is called __________.
4. The relation between angular velocity and frequency of an alternating quantity is given as; ω = __________.
5. In case of sinusoidal voltage; $V_{r.m.s.} = V_m /$ __________.
6. Form factor is defined as the ratio of __________ and __________.

7. One-half cycle of an alternating quantity is called __________.
8. The angular displacement between the positive maximum values of two alternating quantities having same frequency is called __________.
9. Average value of a sinusoidally varying alternating current is given by the relation; $I_{av} = I_m /$ __________.
10. If the frequency of an alternating current is 200 k Hz, its time period will be __________.
11. All the domestic appliances are connected in __________ to the supply.
12. The maximum value of p.f. in an a.c. circuit can be __________.
13. The product of r.m.s. value of voltage and current is called __________.
14. The voltage vector in a pure capacitive circuit will __________ the current vector by 90°.
15. The power drawn by a pure inductor is __________.
16. In an a.c. circuit, the ratio of resistance to impedance is called __________.
17. In a $R - L - C$ series circuit, the current will be maximum when $X_L =$ __________.
18. The consumer having high p.f. will draw __________ current than the consumer having low p.f. for the same load.
19. In a pure inductive circuit, the voltage will __________ the current vector by 90°.
20. At resonance, the p.f. of the circuit is __________.
21. The power consumed in a pure capacitive circuit is __________.
22. The power factor of a pure resistive circuit is __________.
23. The power factor of a pure inductive circuit will be __________

Type-II: Multi-choice questions:

1. *The r.m.s. value an alternating current is given by steady current which when flowing through a given circuit for a given time produces*
   (*a*) the same amount of heat as produced by a.c. when flowing through the same circuit.
   (*b*) more heat than produced by a.c. when flowing through the same circuit
   (*c*) less heat than produced by a.c. when flowing through the same circuit.
   (*d*) 144 calories.
2. *The average value of sinusoidal quantity is given by the relation*
   (*a*) $I_m/\sqrt{2}$ (*b*) 0.707 I_m (*c*) $2\,I_m/\pi$ (*d*) none of above.
3. *In case of an unsymmetrical alternating quantity, the average value must always be taken over*
   (*a*) the half cycle (*b*) the whole cycle
   (*c*) the quarter cycle (*d*) any fraction of the cycle.
4. *The r.m.s. value of sinusoidal alternating current is given by the relation*
   (*a*) $I_m/2$ (*b*) 0.637 I_m (*c*) $2\,I_m/\pi$ (*d*) $I_m/\sqrt{2}$.

5. *The average value of the alternating quantity is more than the r.m.s. value*
(*a*) true (*b*) false

6. *The amplitude factor of sinusoidal current is*
(*a*) 1.11 (*b*) 1.57 (*c*) 1.414 (*d*) 0.637

7. *The r.m.s. value is a fraction of maximum value of sinusoidal current, the value of the fraction is*
(*a*) 1/1.414 (*b*) 0.637 (*c*) $\sqrt{2}$ (*d*) none of above.

8. *The r.m.s. value of an a.c. signal is 10 volts. The peak to peak value will be*
(*a*) 6.37 V (*b*)14.14 V (*c*) 141 V (*d*) 28.28 V

9. *Two waves have the frequency of 600 Hz and one is set at its maximum value whereas the other at zero, the phase angle between them will be*
(*a*) 0° (*b*) 90°
(*c*) 90 × 600/50 degrees (*d*) 90 × 50/600 degrees

10. *A wave completes one cycle in 10μ seconds, its frequency will be*
(*a*) 10μ Hz (*b*) 50 Hz (*c*) 100 k Hz (*d*) 10 k Hz.

11. *An alternating quantity which attains its positive maximum value prior to the other is called the*
(*a*) in phase quantity (*b*) lagging quantity (*c*) leading quantity (*d*) none of above.

12. *If the effective value of the sinusoidal voltage is 111 volts. Its average value will be*
(*a*) 100 V (*b*) 123.21 V (*c*) 156.51 (*d*) none of above

13. *If the frequency of power supply is 60 Hz, the time period of one cycle will be*
(*a*) 0.02 s (*b*) 20 ms (*c*) 16.67 ms (*d*) 0.1667 s.

14. *If a sinusoidal wave has frequency of 50 Hz with 15 ampere r.m.s. value which of the following equation represents this wave*
(*a*) 15 sin 50 *t* (*b*) 30 sin 25 *t* (*c*) 42.42 sin 100 *t* (*d*) 21.21 sin 314 *t*

15. *Peak factor of an alternating current is given by the relation*
(*a*) $I_{r.m.s.}/I_{av}$ (*b*) $I_m/I_{r.m.s.}$ (*c*) $I_{av}/I_{r.m.s.}$ (*d*) $I_{r.m.s.}/I_m$.

16. *The power factor at resonance in R—L—C series circuit is*
(*a*) zero (*b*) unity (*c*) 0.5 lagging (*d*) 0.5 leading

17. *An R—L—C series circuit contains resistance 2.5 Ω, inductance 0.08 H and capacitance of 0.8 μ F, the p.f. of the circuit will be*
(*a*) unity (*b*) zero (*c*) lagging (*d*) leading

18. *The inductive reactance is measured in ohm because it*
(*a*) reduces the magnitude of alternating current
(*b*) increases the magnitude of alternating current
(*c*) is the product of frequency and inductance
(*d*) has a back e.m.f.

19. *When a sinusoidal voltage is applied across an R – L series circuit having R = X_L, the phase angle will be*
(*a*) 90° (*b*) 0° (*c*) 30° (*d*) 45°

20. *If an a.c. voltage is applied across a capacitor, the alternating current flows through the circuit because*
(*a*) of high peak value of voltage.
(*b*) capacitor offers low opposition
(*c*) electrons can pass through capacitor
(*d*) capacitor charges and discharges according to the supply voltage.

21. *At higher frequencies, the value of capacitive reactive*
(*a*) increases
(*b*) decreases
(*c*) remains the same
(*d*) increases and also depends upon applied voltage

22. *In an L-C series circuit, at resonance the*
(*a*) impedance is maximum
(*b*) voltage across C is minimum
(*c*) current is maximum
(*d*) current is minimum

23. *In a pure inductive a.c. circuit:*
(*a*) voltage leads the current vector by 90°
(*b*) current leads the voltage vector by 90°
(*c*) voltage lags the current vector by 90°
(*d*) current is in phase with voltage vector.

24. *What will happen if the frequency of power supply in a pure capacitive circuit is doubled?*
(*a*) the current will also be doubled
(*b*) the current will reduce to half
(*c*) the current will remain the same
(*d*) the current will increase to four fold.

25. *A pure inductance when connected across 230 V, 50 Hz supply, it consumes 40 W. This consumption is*
(*a*) due to reactance of inductor
(*b*) due to current flowing through the inductor
(*c*) due to big size of the inductor
(*d*) none of above; the statement is false.

26. *The power dissipated in the pure capacitance of an R—C series circuit will be*
(*a*) zero
(*b*) small
(*c*) higher than dissipated in resistance
(*d*) equal to dissipated in resistance.

NUMERICALS FOR PRACTICE

1. An alternating voltage is given as $\upsilon = 400 \sin 314\, t$ determine its (*i*) maximum value; (*ii*) effective value; (*iv*) form factor; (*v*) value of voltage after 0.0025 second taking reckoning time from the instant when voltage is zero and becoming positive; (*vi*) time after which voltage attains 200 V for the first time.

(***Ans.*** *400 V; 282.84 V; 50 Hz; 1.11; 282.84 V; 1.67 ms*)

2. An alternating current of frequency 60 Hz has an r.m.s. value of 84.853 A. (*i*) Write down the equation for its instantaneous value; (*ii*) Find the instantaneous value of current 0.002083 second after passing through zero and becoming positive; (*iii*) Find the instantaneous value

of current 1.389 ms after passing through positive maximum value and (*iv*) Time taken to reach 96 A for the first time after passing through positive maximum value.

(P.T.U.) (***Ans.*** *120 sin 377 t; 84.85 A; 103.92 A; 1.7 ms*)

3. For a full wave rectified alternating current, determine (*i*) average value; (*ii*) r.m.s. value; (*iii*) form factor and (*iv*) peak factor. (***Ans.*** $0.637\ I_m$; $0.707\ I_m$; *1.11; 1.4142*)

4. An e.m.f. $e_1 = 50 \sin \omega t$ and the other $e_2 = 30 \sin (\omega t - \pi/6)$ act together in the same circuit. Find the resultant e.m.f. (*i*) analytically and (*ii*) graphically. (***Ans.*** $77.45 \sin (\omega t + 11.17°)$)

5. Three voltages represented by $e_1 = 20 \sin (\omega t + \pi/3)$; $e_2 = 30 \cos (\omega t + \pi/6)$ and $e_3 = 40 \sin (\omega t - \pi/6)$ act together in a circuit. Find an expression for the resultant voltage.

(U.P.T.U.) (***Ans.*** $37.7 \sin (\omega t + 38.17°)$)

6. A 230 V water heater operating for 3 hours daily on a.c. mains is found to give 250 k calorie per hour, the efficiency of heater system being 90%. Determine (*i*) the r.m.s. and the maximum value of current; (*ii*) resistance of the heater; (*iii*) also write down the equations for voltage and current when the supply frequency is 50 Hz.

(***Ans.*** *1.41 A; 1.993 A; 163.1 Ω;* $v = 325.27 \sin 314\, t$; $i = 1.993 \sin 314\, t$)

7. An inductive coil has negligible resistance and inductance of 0.1 henry. It is connected across a 220 V, 50 Hz supply. Find the current and power. Also write down the expression for instantaneous value of applied voltage and current.

(***Ans.*** *7 A; zero;* $v = 311 \sin 314\, t$, $i = 9.9 \sin (314\, t - \pi/2)$)

8. A capacitor has a capacitance of 30 microfarad. Find its capacitive reactance for frequency of 25 and 50 Hz. Find in each case the current if the supply voltage is 400 V.

(***Ans.*** *212.2 Ω; 106.1 Ω; 2.073 A; 4.146 A*)

9. A coil having a resistance of 7 Ω and an inductance of 31.8 m H is connected to 230 V, 50 Hz supply. Calculate (*i*) the circuit current, (*ii*) phase angle, (*iii*) power factor and (*iv*) power consumed. (***Ans.*** *18.85 A; 55° lag; 0.573 lag; 2484.24 W*)

10. A circuit consists of a pure resistance and a coil connected in series. Power dissipated in the resistance and in the coil are 1000 W and 200 W respectively. Voltage drops across the resistance and the coil are 200 V and 300 V respectively. Determine the reactance of the coil and the supply voltage. **(P.T.U.)** (***Ans.*** *49 Ω; 350 V*)

11. A 100 V, 60 W lamp is to be operated at 220 V, 50 Hz mains. What (*i*) pure resistance and (*ii*) pure inductance should be, placed in series with the lamp, which enable it to run without being damaged? Which method would be more economical? (***Ans.*** *200 Ω; 1.038 H*)

12. A resistor *R* in series with a capacitor *C* is connected to a 50 Hz, 230 V supply. Find the value of *C* so that resistance absorbs 500 W at 150 V. Find also the maximum charge and maximum stored energy in C. (**Ans.** $60.85\ \mu F$; 15×10^{-3} *C; 1.85 J*)

13. A coil of resistance 100 Ω and inductance 100 μ H is connected in series with a 100 p F capacitor. The circuit is connected to a 50 V variable frequency supply. Calculate (*i*) the resonant frequency; (*ii*) current at resonance; (*iii*) voltage across *L* and *C* at resonance and (*iv*) *Q*-factor of the circuit. (**Ans.** 1.59×10^6 *Hz; 0.5 A; 500 V; 500 V; 10*)

14. A coil has a resistance of 500 Ω and inductance of 350 μH. Find the capacitance of a capacitor which when connected in parallel with the coil, will produce resonance with a supply frequency of 10^6 Hz. If a second capacitor of capacitance 30 p F is connected in parallel with the first capacitor, find the frequency at which resonance will occur. (**Ans.** *68.8 p F, 0.825 M Hz*)

VIVA-VOCE/REASONING QUESTIONS

1. When a coil is rotated in a magnetic field at a constant speed the wave shape of the induced e.m.f. is sinusoidal (not square or triangular), why?
2. The effective value of current is named as r.m.s. value, why?
3. The r.m.s. value is also called effective value, why?
4. Is the magnitude of voltage supplied to our residences remains only 230 V? tell, why?
5. Why the rms value of an alternating current or voltage is used to denote its magnitude?
6. How are alternating quantities added?
7. Alternating currents are not added arithmetically, why?
8. The power consumed in a pure inductive circuit is zero, why?
9. The power consumed in a pure capacitive circuit is zero, why?
10. Actual power consumed in an a.c. circuit is given by relation $VI \cos \phi$ and not VI, why?
11. Is P.F. more than one, if not why?
12. It is said that the p.f. should be high, why?
13. In ac series circuit at resonant frequency, the value of circuit impedance reduces to resistance, why?

SHORT ANSWER TYPE QUESTIONS

1. What do you understand by alternating current?
2. What is difference between ac and dc?
3. What are the differences between *DC* current and *AC* current.
4. Why sinusoidal voltages and currents are used while generating, transmitting and utilising ac electric power? **(P.T.U., Jan. 2000)**
5. What is meant by instantaneous value?
6. What is meant by peak value of an alternating quantity?
7. What are scalar and vector quantities?
8. What is meant by phase?
9. What is an in-phase condition?
10. What is an out of phase condition?
11. Under what circumstances does a generator produce an alternating induced voltage?
12. List the various factors governing the emf generation?
13. Define average value of an alternating quantity.

14. What is RMS value of an alternating quantity? **(P.T.U., Dec. 2003)**
15. Why are alternating voltages and currents expressed in RMS values instead of average values?
16. Define form factor.
17. Define peak factor of an alternating quantity.
18. What is significance of form factor?
19. What is the significance of peak factor?
20. Difference between form factor and peak factor.
21. What does the form factor of a wave other than sinusoidal wave indicate?
22. Our houses are supplied with alternating voltage whose instantaneous value is given by the equation $e = 325 \sin \omega t$, but we always say that $A.C.$ voltage at home is 230 V. How do you explain this difference?
23. How will you calculate average value for symmetrical waves?
24. What is a phasor?
25. What is the significance of the phasor representation of an alternating quantity?
26. What important points should be considered while representing an alternating quantity by a vector?
27. What is meant by phase difference?
28. What are leading and lagging quantities?
29. A phasor is always assumed to rotate with a constant angular velocity in anticlockwise direction why not clockwise direction?
30. What is an $A.C.$ circuit?
31. What is the phase difference between voltage and current in a pure resistive circuit?
32. What is the average power in a pure resistive circuit?
33. What is the phase difference between current and voltage in an inductor (pure inductance)?
34. Define inductive reactance.
35. What is the instantaneous power in an inductor?
36. What is the value of average power in an inductor?
37. What is the phase difference between voltage and current in a capacitor?
38. Define capacitive reactance.
39. What is the value of average power in a capacitor.
40. What is a series RL circuit?
41. What is the impedance of RL series circuit?
42. What is power factor?
43. What do you understand by RC series circuit?
44. What is the impedance of RC series circuit?
45. What is meant by RLC series circuit?
46. What is the impedance of RLC series circuit?
47. What is mean by resonance circuit?
48. What happens when RLC series circuit is at resonance?

TEST QUESTIONS

1. Explain how a sinusoidal e.m.f. is generated.
2. Derive an expression for the instantaneous value of alternating voltage varying sinusoidally. (P.T.U.)
3. Explain the terms amplitude, cycle, time period, frequency and phase difference as applied to alternating wave forms. (U.P.T.U.)
4. Define r.m.s. value and average value of an alternating quantity.
5. Show that the r.m.s. value of a sinusoidal a.c. voltage of amplitude V_m is $V_m/\sqrt{2}$.
6. Derive the expression for average and r.m.s. value of a sinusoidally varying quantity.
7. Show that the form factor of the sinusoidal wave form is 1.11.
8. Derive the values of form factor and peak factor of a sinusoidally varying quantity.
9. Explain how sinusoidal quantities can be represented by vectors.
10. Explain the component method for addition of alternating quantities.
11. Distinguish between
    (*i*) Time period and frequency
    (*ii*) Cycle and alternation
    (*iii*) Peak factor and form factor
    (*iv*) Average and *rms* value of an alternating current.
12. Explain why the phasor of voltage across an inductor leads its current by 90° and the phasor of voltage across a capacitor lags its current by 90°.
13. Determine phase angle relationship between alternating voltage and current in a purely inductive and a purely capacitance circuit under steady condition.
14. Explain with mathematical expression that power consumed in a pure inductance in zero.
15. Explain with mathematical expression that power consumed in a pure capacitance is zero.
16. Show that power consumed in a pure resistive circuit is not constant but it is fluctuating.
17. Develop the expression for the mean power consumed over a cycle of a single-phase sinusoidal supply delivering power to load comprising of a resistance R in series with an inductance *L*.
18. What is power factor? Discuss the practical importance of p.f.
19. Explain the use of capacitance for p.f. improvement.
20. What is power factor in *R—L—C* series circuit; its leading, lagging and resonance conditions?
21. Define *Q*-factor for the series resonant circuit and express it in terms of the circuit parameters.

ANSWERS TO OBJECTIVE TYPE QUESTIONS

Type-I: Fill in the blanks:

1. max. value; r.m.s. value **2.** r.m.s. **3.** frequency **4.** $2\pi f$
5. $\sqrt{2}$ **6.** r.m.s. value; average value **7.** alternation
8. phase difference **9.** $\pi/2$ **10.** 5μ s. **11.** parallel **12.** one
13. apparent power **14.** lag behind **15.** zero **16.** p.f. **17.** X_C
18. less **19.** lead **20.** unity **21.** zero **22.** unity
23. zero lagging

Type-II: Multi-choice Questions:

1. (*a*) **2.** (*c*) **3.** (*b*) **4.** (*d*) **5.** (*b*)
6. (*c*) **7.** (*a*) **8.** (*d*) **9.** (*b*) **10.** (*c*)
11. (*c*) **12.** (*a*) **13.** (*c*) **14.** (*d*) **15.** (*b*)
16. (*b*) **17.** (*d*) **18.** (*a*) **19.** (*d*) **20.** (*d*)
21. (*b*) **22.** (*c*) **23.** (*a*) **24.** (*a*) **25.** (*d*)
26. (*a*)